Counterexamples
in Analysis

Counterexamples in Analysis

Bernard R. Gelbaum
University of California, Irvine

John M. H. Olmsted
Southern Illinois University

Dover Publications, Inc.
Mineola, New York

Bibliographical Note

This Dover edition, first published in 2003, is an unabridged, slightly corrected republication of the 1965 second printing of the work originally published in
1964 by Holden-Day, Inc., San Francisco.

Library of Congress Cataloging-in-Publication Data

Gelbaum, Bernard R.
    Counterexamples in analysis / Bernard R. Gelbaum, John M.H. Olmsted.
        p. cm.
    " . . . an unabridged, slightly corrected republication of the 1965 second
printing of the work originally published in 1964 by Holden-Day, San
Francisco [in the Mathesis series]"--T.p. verso.
    Includes bibliographical references and index.
    ISBN-13: 978-0-486-42875-8
    ISBN-10: 0-486-42875-3
    1. Mathematical analysis. I. Olmsted, John Meigs Hubbell, 1911- II. Title.

QA300.G4 2003
515--dc21
                                                            2002041784

Manufactured in the United States by Courier Corporation
42875306
www.doverpublications.com

Preface

The principal type of question asked in mathematics is, "Is statement S true?" where the statement S is of the form "Every member of the class A is a member of the class $B: A \subset B$." To demonstrate that such a statement is *true* means to formulate a proof of the inclusion $A \subset B$. To demonstrate that such a statement is *false* means to find a member of A that is *not* a member of B, in other words a *counterexample*. To illustrate, if the statement S is "Every continuous function is somewhere differentiable," then the sets A and B consist of all continuous functions and all functions that are somewhere differentiable, respectively; Weierstrass's celebrated example of a function f that is continuous but nowhere differentiable (cf. Example 8, Chapter 3) is a counterexample to the inclusion $A \subset B$, since f is a member of A that is not a member of B. At the risk of oversimplification, we might say that (aside from definitions, statements, and hard work) mathematics consists of two classes—proofs and counterexamples, and that mathematical discovery is directed toward two major goals—the formulation of proofs and the construction of counterexamples. Most mathematical books concentrate on the first class, the body of proofs of true statements. In the present volume we address ourselves to the second class of mathematical objects, the counterexamples for false statements.

Generally speaking, examples in mathematics are of two types, illustrative examples and counterexamples, that is, examples to show why something makes sense and examples to show why something does *not* make sense. It might be claimed that *any* example is a counterexample to *something*, namely, the statement that such an example is impossible. We do not wish to grant such universal interpretation to the term *counterexample*, but we do suggest that its meaning be sufficiently broad to include any example whose role is not that of illustrating a true theorem. For instance, a polynomial as an example of a continuous function is *not* a counterexample, but a polynomial as an example of a function that fails to be bounded or of a function that fails to be periodic *is* a counterexample. Similarly, the class of all monotonic functions on a bounded closed interval as a class of integrable functions is *not* a counterexample, but this same class as an example of a function space that is not a vector space *is* a counterexample.

The audience for whom this book is intended is broad and varied. Much of the material is suitable for students who have not yet completed a first course in calculus, and for teachers who may wish to make use of examples to show to what extent things may "go wrong" in calculus. More advanced students of analysis will discover nuances that are usually by-passed in standard courses. Graduate students preparing for their degree examinations will be able to add to their store of important examples delimiting the range of the theorems they have learned. We hope that even mature experts will find some of the reading new and worthwhile.

The counterexamples presented herein are limited almost entirely to the part of analysis known as "real variables," starting at the level of calculus, although a few examples from metric and topological spaces, and some using complex numbers, are included. We make no claim to completeness. Indeed, it is likely that many readers will find some of their favorite examples missing from this collection, which we confess is made up of *our* favorites. Some omissions are deliberate, either because of space or because of favoritism. Other omissions will undoubtedly be deeply regretted when they are called to our attention.

This book is meant primarily for browsing, although it should be a useful supplement to several types of standard courses. If a reader finds parts hard going, he should skip around and pick up something new and stimulating elsewhere in the book. An attempt has been made to grade the contents according to difficulty or sophistication within the following general categories: (i) the chapters, (ii) the topics within chapters, and (iii) the examples within topics. Some knowledge of related material is assumed on the part of the reader, and therefore only a minimum of exposition is provided. Each chapter is begun with an introduction that fixes notation, terminology, and definitions, and gives statements of some of the more important relevant theorems. A substantial bibliography is included in the back of the book, and frequent reference is made to the articles and books listed there. These references are designed both to guide the reader in finding further information on various subjects, and to give proper credits and source citations. If due recognition for the authorship of any counterexample is lacking, we extend our apology. Any such omission is unintentional.

Finally, we hope that the readers of this book will find both enjoyment and stimulation from this collection, as we have. It has been our experience that a mathematical question resolved by a counterexample has the pungency of good drama. Many of the most elegant and artistic contributions to mathematics belong to this genre.

B.R.G.

Irvine, California

J.M.H.O.

Carbondale, Illinois

Table of Contents

2. Functions and Limits

3. Differentiation

4. Riemann Integration

6. Infinite Series

7. Uniform Convergence

8. Sets and Measure on the Real Axis

Part II. Higher Dimensions

9. Functions of Two Variables

10. Plane Sets

11. Area

12. Metric and Topological Spaces

13. Function Spaces

Part I
Functions of a Real Variable

Chapter 1
The Real Number System

Introduction

We begin by presenting some definitions and notations that are basic to analysis and essential to this first chapter. These will be given in abbreviated form with a minimum of explanatory discussion. For a more detailed treatment see [16], [21], [22], and [30] of the Bibliography.

If A is any set of objects, the statement *a is a member of A* is written $a \in A$. The contrary statement that a is *not* a member of A is written $a \notin A$. If A and B are sets, the statement *A is a subset of B* is written $A \subset B$, and is equivalent to the implication $x \in A$ *implies* $x \in B$, also written $x \in A \Rightarrow x \in B$. The phrase *if and only if* is often abbreviated iff, and sometimes symbolized $\Leftrightarrow$. The set whose members are $a, b, c, \cdots$ is denoted $\{a, b, c, \cdots\}$. The notation $\{\cdots \mid \cdots\}$ is used to represent the set whose general member is written between the first brace { and the vertical bar |, and whose defining property or properties are written between the vertical bar | and the second brace }. The **union** and **intersection** of the two sets A and B can therefore be defined:

$$A \cup B \equiv \{x \mid x \in A \quad \text{or} \quad x \in B\},$$

$$A \cap B \equiv \{x \mid x \in A, x \in B\},$$

where the comma in the last formula stands for *and*. For convenience, members of sets will often be called *points*. The **difference** between the sets A and B is denoted $A \setminus B$ and defined:

$$A \setminus B \equiv \{x \mid x \in A, x \notin B\}.$$

When a general *containing set* or *space* or *universe of discourse* S is clearly indicated or understood from context, the difference $S \setminus A$ is called the **complement** of A, and denoted A'. In general, the difference $A \setminus B$ is called the **complement of B relative to A**.

If A and B are two nonempty sets (neither A nor B is the empty set $\emptyset$), their **Cartesian product** is the set of all ordered pairs (a, b), where $a \in A$ and $b \in B$, denoted:

$$A \times B \equiv \{(a, b) \mid a \in A, b \in B\}.$$

If $(a, b) \in A \times B$, a is the **first coordinate** or **component** of (a, b) and b is the **second coordinate** or **component** of (a, b). Any subset ρ of $A \times B$ is called a **relation from A to B**. A **function** from A to B is a relation f from A to B such that no two distinct members of f have the same first coordinate. If the phrases *there exists* and *there exist* are symbolized by the **existential quantifier $\exists$**, and the words *such that* by the symbol $\ni$, the **domain** (of definition) $D = D_f$ and **range** (of values) $R = R_f$ of a function f can be defined:

$$D = D_f \equiv \{x \mid \exists y \ni (x, y) \in f\},$$

$$R = R_f \equiv \{y \mid \exists x \ni (x, y) \in f\}.$$

The function f is a **function on A into B** iff f is a function from A to B with domain equal to A. The function f is a **function on A onto B** iff f is a function on A into B with range equal to B. A function f is a **one-to-one correspondence** between the members of A and the members of B iff f is a function on A onto B such that no two distinct members of f have the same second coordinate. The **values** of a function are the members of its range. The inverse f^{-1} of a one-to-one correspondence f is obtained by interchanging the domain and range of f:

$$f^{-1} = \{(x, y) \mid (y, x) \in f\}.$$

A **constant function** is a function whose range consists of one point.

Various types of relations and functions are indicated in Figure 1. In each case the sets A and B are taken to be the closed unit interval $[0, 1]$ consisting of all real numbers x such that $0 \leq x \leq 1$.

Let f be a function on A into B, symbolized in the following two

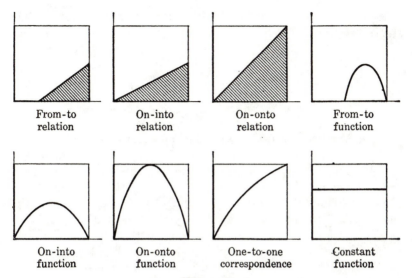

| From-to relation | On-into relation | On-onto relation | From-to function |
| On-into function | On-onto function | One-to-one correspondence | Constant function |

Figure 1

ways:

$$f : A \to B,$$

$$A \xrightarrow{\ f\ } B.$$

If x is an arbitrary member of A, then there is exactly one member y of B such that $(x, y) \in f$. This member y of B is written:

$$y = f(x).$$

Other ways of writing the function just described are:

$$f : y = f(x), \quad x \in A, \quad y \in B;$$

$$f : x \in A, \quad f(x) \in B;$$

or,

$$y = f(x) : x \in A, \quad y \in B,$$

$$f(x) : x \in A,$$

when it is clear from the context that the notation $f(x)$ represents a *function* rather than merely one of its values.

5

If f is a function with domain D, and if S is a subset of D, then the **restriction** of f to S is the function g whose domain is S such that

$$x \in S \Rightarrow g(x) = f(x).$$

The range of the restriction of f to S is denoted $f(S)$. That is,

$$f(S) = \{y \mid \exists\, x \in S \ni f(x) = y\}.$$

If g is a restriction of f, then f is called an **extension** of g.

If f and g are functions such that the range of g is a subset of the domain of f, then the **composite** of f and g, denoted $f \circ g$, is the function whose value at the point x of the domain of g is $f(g(x))$; in short, the *function $y = f(g(x))$* is called the composite of the *function $f(u)$* and the *function $u = g(x)$*. (It should be noted that the composite of f and g is not in general the same as the composite of g and f; counterexample: $(x + 1)^2 \neq x^2 + 1$.)

If A is a nonempty set, a **binary operation from A to A** is a function from $A \times A$ to A. A **binary operation on A into A** is a function on $A \times A$ into A. In classical arithmetic there are two fundamental binary operations: addition and multiplication. Many properties of these operations in arithmetic are shared by operations in more abstract settings, where the operations bear the same names. If a binary operation F is called **addition**, and if $z = F((x, y))$, then z is also written $z = x + y$. If a binary operation G is called **multiplication**, and if $z = G((x, y))$, then z is also written $z = xy$, or $z = x \cdot y$.

Definition I. *A field is a nonempty set $\mathfrak{F}$, together with two binary operations on $\mathfrak{F}$ into $\mathfrak{F}$, called addition and multiplication, such that:*
A. For addition:
  *(i) The **associative law** holds:*
$$x, y, z \in \mathfrak{F} \Rightarrow x + (y + z) = (x + y) + z.$$
  (ii) $\exists$ a member 0 of $\mathfrak{F}$ such that
$$x \in \mathfrak{F} \Rightarrow x + 0 = x.$$
  (iii) $x \in \mathfrak{F} \Rightarrow \exists\, (-x) \in \mathfrak{F} \ni x + (-x) = 0.$
  *(iv) The **commutative law** holds:*
$$x, y \in \mathfrak{F} \Rightarrow x + y = y + x.$$

B. For multiplication:

   (*i*) *The **associative law** holds:*
$$x, y, z \in \mathfrak{F} \Rightarrow x(yz) = (xy)z.$$

   (*ii*) $\exists$ *a member 1 of $\mathfrak{F}$ such that* $1 \neq 0$ *and* $x \in \mathfrak{F} \Rightarrow x \cdot 1 = x.$

   (*iii*) $x \in \mathfrak{F}, x \neq 0 \Rightarrow \exists \ x^{-1} \in \mathfrak{F} \ni x \cdot x^{-1} = 1.$

   (*iv*) *The **commutative law** holds:*
$$x, y \in \mathfrak{F} \Rightarrow xy = yx.$$

C. For addition and multiplication:

   *The **distributive law** holds (more precisely, multiplication is distributive over addition):*

$$x, y, z \in \mathfrak{F} \Rightarrow x(y + z) = xy + xz.$$

The member 0 of $\mathfrak{F}$, of A(*ii*), is called the **zero**, or **additive identity**, of $\mathfrak{F}$. The member $(-x)$ of $\mathfrak{F}$, of A(*iii*), is called the **negative**, or **additive inverse**, of x. The binary operation $x - y$, defined by $x - y \equiv x + (-y)$, is called **subtraction**. The member 1 of $\mathfrak{F}$, of B(*ii*), is called the **one**, or **unity**, or **multiplicative identity**, of $\mathfrak{F}$. The member x^{-1} of $\mathfrak{F}$, of B(*iii*), is called the **reciprocal**, or **multiplicative inverse**, of x. The binary operation x/y, defined by $x/y \equiv xy^{-1}$, where $y \neq 0$, is called **division**. Division is a "from-to" operation and not an "on-into" operation since "division by zero" is excluded.

A nonempty set $\mathcal{G}$ together with a binary operation on $\mathcal{G}$ into $\mathcal{G}$ — in this case denoted $+$ and called addition — subject to properties A(*i*), (*ii*), and (*iii*) is called a **group** (in this case, an *additive* group). In case the commutative law A(*iv*) holds, $\mathcal{G}$ is called an **Abelian** or **commutative** group. Thus, with respect to addition any field is an Abelian additive group. With respect to multiplication the nonzero members of a field form an Abelian *multiplicative* group.

Definition II. *An **ordered field** is a field $\mathfrak{F}$ that contains a subset $\mathcal{P}$ such that*

   (*i*) *$\mathcal{P}$ is closed with respect to addition; that is,*

$$x \in \mathcal{P}, y \in \mathcal{P} \Rightarrow x + y \in \mathcal{P}.$$

   (*ii*) *$\mathcal{P}$ is closed with respect to multiplication; that is,*

$$x \in \mathcal{P}, y \in \mathcal{P} \Rightarrow xy \in \mathcal{P}.$$

(*iii*) $x \in \mathfrak{F} \Rightarrow$ *exactly one of the three statements is true:*

$$x \in \mathcal{P}; \qquad x = 0; \qquad -x \in \mathcal{P}.$$

A member x of $\mathfrak{F}$ is **positive** *iff $x \in \mathcal{P}$; x is* **negative** *iff $-x \in \mathcal{P}$.*

Inequalities in an ordered field are defined in terms of membership in $\mathcal{P}$. For example,

$$x < y \quad \text{iff} \quad y - x \in \mathcal{P};$$

$$x \geqq y \quad \text{iff} \quad x - y \in \mathcal{P} \quad \text{or} \quad x = y.$$

A function f from $\mathfrak{F}$ to $\mathfrak{F}$, where $\mathfrak{F}$ is an ordered field, is **increasing** (or **nondecreasing**) on a subset A of its domain iff

$$x, y \in A, x < y \Rightarrow f(x) \leqq f(y).$$

The function f is **strictly increasing** on A iff

$$x, y \in A, x < y \Rightarrow f(x) < f(y).$$

The terms **decreasing** (or **nonincreasing**) and **strictly decreasing** are similarly defined. A function is **monotonic** on a set iff it is either increasing or decreasing there. **Strictly monotonic** has an obvious definition.

If $\mathfrak{F}$ is an ordered field and if $x \in \mathfrak{F}$, then $|x|$, called the **absolute value** of x, is defined to be x in case $x \geqq 0$, and to be $-x$ in case $x < 0$.

The following are a few of the standard properties of absolute value, where $x, y, \varepsilon \in \mathfrak{F}$.

(*i*) $|x| \geqq 0; |x| = 0$ iff $x = 0$.

(*ii*) $|xy| = |x| \cdot |y|$.

(*iii*) If $\varepsilon > 0$, $|x| < \varepsilon$ iff $-\varepsilon < x < \varepsilon$.

(*iv*) The **triangle inequality**: $|x + y| \leqq |x| + |y|$.

(*v*) $|x| = \sqrt{x^2}$; that is, $|x|$ is the unique member of $\mathcal{P} \cup \{0\}$ whose square $= x^2$.

(*vi*) $||x| - |y|| \leqq |x - y|$.

If $\mathfrak{F}$ is an ordered field, and if $a, b \in \mathfrak{F}$, $a < b$, then the following sets are called **finite** or **bounded** intervals, further described by the attached initial adjective, and denoted as indicated with parentheses and/or brackets:

open: $\qquad\qquad\qquad\qquad (a, b) \equiv \{x \mid x \in \mathfrak{F}, \quad a < x < b\},$

closed: $\qquad\qquad\qquad\quad [a, b] \equiv \{x \mid x \in \mathfrak{F}, \quad a \leqq x \leqq b\},$

half-open (or half-closed): $\quad [a, b) \equiv \{x \mid x \in \mathfrak{F}, \quad a \leqq x < b\},$

half-open (or half-closed): $\quad (a, b] \equiv \{x \mid x \in \mathfrak{F}, \quad a < x \leqq b\}.$

Infinite or **unbounded** intervals are similarly delineated:

open: $\qquad\qquad (a, +\infty) \equiv \{x \mid x > a\},$

open: $\qquad\qquad (-\infty, a) \equiv \{x \mid x < a\},$

closed: $\qquad\qquad [a, +\infty) \equiv \{x \mid x \geqq a\}.$

closed: $\qquad\qquad (-\infty, a] \equiv \{x \mid x \leqq a\}.$

open and closed: $\quad (-\infty, +\infty) \equiv \mathfrak{F}.$

A **neighborhood** of a point a of an ordered field $\mathfrak{F}$ is an open interval of the form $(a - \varepsilon, a + \varepsilon)$, where ε is a positive member of $\mathfrak{F}$. This neighborhood can also be written in terms of absolute values, and will be denoted:

$$N(a, \varepsilon) \equiv (a - \varepsilon, a + \varepsilon) \equiv \{x \mid |x - a| < \varepsilon\}.$$

A **deleted neighborhood** of a point a is a neighborhood of a with the point a deleted; a deleted neighborhood $D(a, \varepsilon)$ of a, for some $\varepsilon > 0$, is thus defined:

$$D(a, \varepsilon) \equiv \{x \mid 0 < |x - a| < \varepsilon\}.$$

The binary operations max and min on $\mathfrak{F}$ to $\mathfrak{F}$ are defined:

$$\max (x, y) \equiv \begin{cases} x & \text{if} \quad x \geqq y, \\ y & \text{if} \quad x < y; \end{cases} \qquad \min (x, y) \equiv \begin{cases} y & \text{if} \quad x \geqq y, \\ x & \text{if} \quad x < y. \end{cases}$$

If $\mathfrak{F}$ is an ordered field, if $u \in \mathfrak{F}$, and if $x \leqq u$ for every member x of a nonempty set A of points of $\mathfrak{F}$, then u is called an **upper bound** of A. A nonempty set in $\mathfrak{F}$ is **bounded above** in $\mathfrak{F}$ iff there exists a member of $\mathfrak{F}$ that is an upper bound of the set. If s is an upper bound of A and if s is less than every other upper bound of A, then s is called the **least upper bound** or **supremum** of A, written $s =$

sup (A) = sup A. Similar definitions hold for **lower bound, bounded below,** and **greatest lower bound** or **infimum** i of a nonempty set A, written i = inf (A) = inf A.

Definition III. *A* **complete ordered field** *is an ordered field* $\mathfrak{F}$ *in which a least upper bound exists for every nonempty set in* $\mathfrak{F}$ *that is bounded above in* $\mathfrak{F}$.

Any two complete ordered fields $\mathfrak{F}$ and $\mathfrak{F}'$ are isomorphic in the sense that there exists a one-to-one correspondence $x \leftrightarrow x'$, where $x \in \mathfrak{F}$ and $x' \in \mathfrak{F}'$, that preserves binary operations and order; that is:

$$(x + y)' = x' + y', \quad (xy)' = x'y', \quad x < y \text{ iff } x' < y'.$$

(For a proof and discussion, see [35], pp. 128–131.) As far as *structure* is concerned, then, the real number system is uniquely described by the definition:

Definition IV. *The* **real number system** $\mathfrak{R}$ *is a complete ordered field.*

A function on A onto B is called **real-valued** iff $B \subset \mathfrak{R}$; it is called **a function of a real variable** iff $A \subset \mathfrak{R}$.

The **signum function** is the real-valued function of a real variable defined and denoted: sgn $x \equiv 1$ if $x > 0$; sgn $x \equiv -1$ if $x < 0$; sgn $0 \equiv 0$.

If S is any nonempty space and if A is any subset of S, then the **characteristic function** of A is the real-valued function χ_A defined: $\chi_A(x) \equiv 1$ if $x \in A$ and $\chi_A(x) \equiv 0$ if $x \in A' = S \setminus A$.

Definition V. *An* **inductive set** *in an ordered field* $\mathfrak{F}$ *is a set A having the two properties:*
  (*i*) $1 \in A$.
  (*ii*) $x \in A \Rightarrow x + 1 \in A$.

Definition VI. *A member n of an ordered field* $\mathfrak{F}$ *is a* **natural number** *iff n is a member of every inductive set of* $\mathfrak{F}$. *The set of all natural numbers of* $\mathfrak{F}$ *is denoted* $\mathfrak{N}$.

From this definition follow the familiar properties (cf. [35] pp. 17–18) of natural numbers, including the theorem:

Fundamental Theorem of Induction. If $\mathcal{S}$ is an inductive set of natural numbers, then $\mathcal{S} = \mathfrak{N}$.

If $\mathfrak{N}$ and $\mathfrak{N}^*$ are the sets of all natural numbers of two ordered fields $\mathfrak{F}$ and $\mathfrak{F}^*$, then $\mathfrak{N}$ and $\mathfrak{N}^*$ are isomorphic (cf. [35], pp. 34–35).

Definition VII. *A member x of an ordered field $\mathfrak{F}$ is an **integer** iff $x \in \mathfrak{N}$, $x = 0$, or $-x \in \mathfrak{N}$. A member x of an ordered field is a **rational number** iff there exist integers m and n, $n \neq 0$, such that $x = m/n$.*

The set of all rational numbers of an ordered field $\mathfrak{F}$, under the operations of addition and multiplication of $\mathfrak{F}$ and the ordering of $\mathfrak{F}$, is also an ordered field, denoted $\mathcal{Q}$. (Any two ordered fields of rational numbers are isomorphic; cf. [35], p. 67.)

Definition VIII. *A **ring** is a nonempty set $\mathcal{B}$, together with two binary operations on $\mathcal{B}$ into $\mathcal{B}$, called **addition** and **multiplication**, such that the following laws of Definition I hold: A(i), (ii), (iii), (iv), B(i), C, and a second distributive law:*

C′. $\qquad x, y, z \in \mathcal{B} \Rightarrow (x + y)z = xz + yz$.

Definition IX. *An **integral domain** is a ring $\mathfrak{D}$ such that the following additional laws of Definition I hold: B(ii), (iv) — that is, all laws of Definition I except for B(iii) — and also the following weakened form of B(iii):*

D. $\qquad x \in \mathfrak{D}, y \in \mathfrak{D}, x \neq 0, y \neq 0 \Rightarrow xy \neq 0$.

That D is a weakened form (that is, a consequence) of B(iii) can be seen by assuming the existence of $x \neq 0$ and $y \neq 0$ such that $xy = 0$. Then $x^{-1}(xy) = (x^{-1}x)y = 1y = y \neq 0$, whereas $x^{-1}0 = 0$. (Contradiction.) In any ring, law D is equivalent to the law:

D′. Cancellation law. $\quad xy = xz, x \neq 0 \Rightarrow y = z$.

($D \Rightarrow D'$ since $xy = xz$ iff $x(y - z) = 0$; $D' \Rightarrow D$ since $xy = 0$ can be written $xy = x0$.)

The set of all integers of an ordered field $\mathfrak{F}$, with the operations of addition and multiplication of $\mathfrak{F}$, is an integral domain, denoted $\mathcal{J}$. Any two integral domains of integers are isomorphic (cf. [35], p. 64).

Let f be a function from $\mathfrak{F}$ to $\mathfrak{F}$, where $\mathfrak{F}$ is an ordered field, and let $a \in \mathcal{F}$. Then f is **continuous** at a iff a belongs to the domain D of f, and corresponding to an arbitrary positive member ε of $\mathcal{F}$ $\exists$ a positive member δ of $\mathcal{F}$ such that $|f(x) - f(a)| < \varepsilon$ for every x of D such that $|x - a| < \delta$. With the aid of the **universal quantifier** $\forall$, representing the words *for all, for an arbitrary, for every,* or *for each,* and the language and notation of neighborhoods, this last portion of the definition of continuity of f at a point a of D can be expressed:

$$\forall \varepsilon > 0 \, \exists \, \delta > 0 \ni f(D \cap N(a, \delta)) \subset N(f(a), \varepsilon).$$

A point p is a **limit point** of a nonempty set A, in an ordered field $\mathcal{F}$, iff every deleted neighborhood of p contains at least one point of A :

$$\forall \varepsilon > 0 \, \exists \, a \in D(p, \varepsilon) \cap A.$$

If f is a function from $\mathcal{F}$ to $\mathcal{F}$, if a is a limit point of the domain D of f, and if $b \in \mathcal{F}$, then the **limit** of $f(x)$ as x approaches a is said to exist and equal b, with the notation

$$\lim_{x \to a} f(x) = b,$$

iff

$$\forall \varepsilon > 0 \, \exists \, \delta > 0 \ni f(D \cap D(a, \delta)) \subset N(b, \varepsilon).$$

One-sided limits are defined similarly, and denoted $\lim_{x \to a+} f(x)$ and $\lim_{x \to a-} f(x)$.

A function f from an ordered field $\mathcal{F}$ to $\mathcal{F}$ is **uniformly continuous** on a subset A of its domain D iff

$$\forall \varepsilon > 0 \, \exists \, \delta > 0 \ni$$

$$x_{,1} \, x_2 \in A, |x_1 - x_2| < \delta \Rightarrow |f(x_1) - f(x_2)| < \varepsilon.$$

If f is a function from an ordered field $\mathcal{F}$ to $\mathcal{F}$, and if a is a point of the domain D of f, then the symbol $f'(a)$ denotes the member of $\mathcal{F}$ defined

$$f'(a) \equiv \lim_{x \to a} \frac{f(x) - f(a)}{x - a} = \lim_{h \to 0} \frac{f(a + h) - f(a)}{h},$$

provided this limit exists. The function f' defined by $f'(x)$ whenever $f'(x)$ exists for x in the domain of D is called the **derivative** of f.

A function f from an ordered field $\mathfrak{F}$ to $\mathfrak{F}$ is said to have the **intermediate value property** on an interval I contained in its domain iff

$$\forall\, a, b \in I,\, d \in \mathfrak{F} \ni a < b$$

$$\text{and either } f(a) < d < f(b) \text{ or} f(a) > d > f(b),$$

$$\exists\, c \ni a < c < b, \qquad f(c) = d.$$

A **sequence** is a function whose domain is the set of natural numbers $\mathfrak{N}$. Its value for n is usually denoted by means of a subscript, thus: a_n, and the sequence itself by braces: $\{a_n\}$. A sequence $\{a_n\}$, where the values or **terms** a_n are members of an ordered field $\mathfrak{F}$, is said to be **convergent** and to have the **limit** b, or to **converge** to b, where $b \in \mathfrak{F}$, iff

$$\forall\, \varepsilon \in \mathcal{P} \,\exists\, N \in \mathfrak{N} \ni n > N \Rightarrow |\, a_n - b\,| < \varepsilon,$$

where $\mathcal{P}$ is the set of positive members of $\mathfrak{F}$. A sequence is **divergent** iff it fails to be convergent (that is, iff a limit b fails to exist). A sequence $\{a_n\}$, where the terms a_n are members of an ordered field $\mathfrak{F}$, is a **Cauchy sequence** iff

$$\forall\, \varepsilon \in \mathcal{P} \,\exists\, N \in \mathfrak{N} \ni m, n > N \Rightarrow |\, a_m - a_n\,| < \varepsilon.$$

Every convergent sequence is a Cauchy sequence, and if $\mathfrak{F} = \mathfrak{R}$ every Cauchy sequence is convergent (cf. [34], p. 57).

A **complex number** is an ordered pair (x, y) of real numbers x and y. **Addition** and **multiplication** of complex numbers are defined:

$$(x, y) + (u, v) \equiv (x + u, y + v),$$

$$(x, y)(u, v) \equiv (xu - yv, xv + yu).$$

The complex numbers form a field $\mathcal{C}$ (cf. [34], p. 497), with zero $(0, 0)$ and unity $(1, 0)$. In the sequel the standard notation $x + iy$ for the ordered pair (x, y) will be usual.

1. An infinite field that cannot be ordered.

To say that a field $\mathfrak{F}$ cannot be ordered is to say that it possesses no subset ℙ satisfying the three properties of Definition II of the Introduction. A preliminary comment is that since every ordered field is infinite, no finite field can be ordered ([35], p. 38).

An example of an *infinite* field that cannot be ordered is the field ℂ of complex numbers. To show that this is the case, assume that there does exist a subset ℙ of ℂ satisfying Definition II. Consider the number $i \equiv (0, 1)$. Since $i \neq (0, 0)$, there are two alternative possibilities. The first is $i \in ℙ$, in which case $i^2 = (-1, 0) \in ℙ$, whence $i^4 = (1, 0) \in ℙ$. Since i^2 and i^4 are additive inverses of each other, and since it is impossible for two additive inverses both to belong to ℙ (cf. Definition II, (*iii*)), we have obtained a contradiction, as desired. The other option is $-i = (0, -1) \in ℙ$, in which case $(-i)^2 = (-1, 0) \in ℙ$, whence $(-i)^4 = (1, 0) \in ℙ$, with the same contradiction as before.

2. A field that is an ordered field in two distinct ways.

The set 𝔉 of all numbers of the form $r + s\sqrt{2}$, where r and s are rational and the operations of addition and multiplication are those of the real number system ℜ of which 𝔉 is a subset, is an ordered field in which the subset ℙ of Definition II is the set of all members of 𝔉 that are positive members of ℜ, that is, positive real numbers. A second way in which 𝔉 is an ordered field is provided by the subset ℬ defined:

$$r + s\sqrt{2} \in ℬ \Leftrightarrow r - s\sqrt{2} \in ℙ.$$

That ℬ satisfies the three requirements of Definition II is easily verified.

Each of the fields ℚ of rational numbers and ℜ of real numbers is an ordered field in only one way ([35], p. 146).

3. An ordered field that is not complete.

The ordered field ℚ of rational numbers is not complete. This can be seen as follows: The set A of all positive rational numbers whose squares are less than 2,

$$A \equiv \{r \mid r \in ℚ, \quad r > 0, \quad r^2 < 2\},$$

is nonempty ($1 \in A$) and is bounded above by the rational number 2. Let us assume that ℚ is complete. Then there must be a positive rational number c that is the supremum of A. Since there is no rational number whose square is equal to 2 (cf. [35], p. 126), either $c^2 < 2$ or $c^2 > 2$. Assume first that $c^2 < 2$ and let d be the positive number

$$d = \frac{1}{2} \min \left(\frac{2 - c^2}{(c + 1)^2}, 1 \right).$$

Then $c + d$ is a positive rational number greater than c whose square is less than 2:

$$(c + d)^2 < c^2 + d(c + 1)^2 < 2.$$

But this means that $c + d \in A$, whereas c is an upper bound of A. (Contradiction.) Now assume that $c^2 > 2$ and let d be the positive number

$$d = \frac{c^2 - 2}{2(c + 1)^2}.$$

Then $c - d$ is a positive rational number less than c whose square is greater than 2:

$$(c - d)^2 > c^2 - d(c + 1)^2 > 2.$$

Since $c - d$ is therefore an upper bound of A less than the *least* upper bound c, a final contradiction is reached.

4. A non-Archimedean ordered field.

An ordered field $\mathfrak{F}$ is **Archimedean** iff the set $\mathfrak{N}$ of natural numbers of $\mathfrak{F}$ is not bounded above in $\mathfrak{F}$ (equivalently, whenever a, $b \in \mathfrak{F}$, $a > 0, b > 0$, then there exists a natural number n such that $na > b$). Let f be a **polynomial function** on $\mathfrak{R}$ into $\mathfrak{R}$:

$$f(x) = \sum_{k=0}^{n} a_k x^k, \qquad a_k \in \mathfrak{R}, \quad k = 0, 1, \cdots, n,$$

and let g be a nonzero polynomial function (that is, $g(x)$ is not *identically* zero), and let f/g be the rational function h defined by $h(x) = f(x)/g(x)$ whose domain consists of all real numbers for which $g(x) \neq 0$. Let $\mathfrak{X}$ consist of all rational functions f/g in lowest terms (the only common polynomial factors of f and g are constants), with addition and multiplication defined:

$$\frac{f}{g} + \frac{r}{s} \equiv \frac{fs + gr}{gs}, \qquad \frac{f}{g} \cdot \frac{r}{s} \equiv \frac{fr}{gs},$$

where the right-hand member in each case is reduced to lowest terms. Then $\mathfrak{IC}$ is a field ([35], p. 104). If a subset $\mathcal{P}$ of $\mathfrak{IC}$ is defined to consist of all nonzero f/g of $\mathfrak{IC}$ such that the leading coefficients (that is, the coefficients of the terms of highest degree) of f and g have the same sign, then $\mathcal{P}$ satisfies the requirements of Definition II, and $\mathfrak{IC}$ is an ordered field. But any rational function $f/1$, where f is a nonconstant polynomial with positive leading coefficient, is an upper bound of the set $\mathfrak{N}$ of natural numbers of $\mathfrak{IC}$ (the natural numbers of $\mathfrak{IC}$ are the constant rational functions of the form $n/1$, where n is the constant polynomial whose values are all equal to the real natural number n). For a more detailed discussion, see [35], pp. 99–108.

5. An ordered field that cannot be completed.

To say that an ordered field $\mathfrak{F}$ cannot be completed means that there is no complete ordered field $\mathfrak{R}$ containing $\mathfrak{F}$ in such a way that the operations of addition and multiplication and the order relation of $\mathfrak{F}$ are consistent with those of $\mathfrak{R}$. The preceding example $\mathfrak{IC}$ of rational functions cannot be completed in this sense or, in other words, cannot be *embedded* in the real number system (cf. Definition IV). The reason, in brief, is that if $\mathfrak{IC}$ *could* be embedded in $\mathfrak{R}$, then the natural numbers of $\mathfrak{IC}$ would correspond in an obvious fashion with those of $\mathfrak{R}$. Since $\mathfrak{N}$ is bounded above in $\mathfrak{IC}$ but not in $\mathfrak{R}$ ([35], p. 122), a contradiction is obtained.

6. An ordered field where the rational numbers are not dense.

The "rational numbers" of the ordered field $\mathfrak{IC}$ of Example 4 are not dense in $\mathfrak{IC}$. That is, there are two distinct members of $\mathfrak{IC}$ having no rational number between them. In fact, any ordered field $\mathfrak{F}$ in which the rational numbers are dense is Archimedean. To see this, let a be an arbitrary positive member of $\mathfrak{F}$, and let m/n be a rational number between 0 and $1/a$. Adjust notation if necessary in order to assume (without loss of generality) that m and n are both positive.

Then

$$0 < \frac{1}{n} \leq \frac{m}{n} < \frac{1}{a},$$

whence $n > a$. Consequently a is not an upper bound of $\mathfrak{N}$, and since a is arbitrary, $\mathfrak{N}$ is not bounded above. It follows, then, that since $\mathfrak{IC}$ is *not* Archimedean the rational numbers of $\mathfrak{IC}$ *cannot* be dense in $\mathfrak{IC}$. Examples of two distinct members of $\mathfrak{IC}$ having no rational number between them are any two distinct nonconstant polynomials with positive leading coefficients.

7. An ordered field that is Cauchy-complete but not complete.

If the ordered field $\mathfrak{IC}$ of rational functions, Example 4, is extended by means of equivalence classes of Cauchy sequences, the resulting structure is an ordered field in which every Cauchy sequence converges. However, by Example 5, this Cauchy-completion cannot be complete in the sense of the definition given in the Introduction in terms of least upper bounds. (For a treatment of Cauchy-completion in general, see [20], pp. 106–107, [21].)

8. An integral domain without unique factorization.

A **unit** of an integral domain $\mathfrak{D}$ is a member u of $\mathfrak{D}$ having a multiplicative inverse v in $\mathfrak{D} : uv = 1$. (The units of the integral domain $\mathfrak{g}$ of integers are 1 and -1.) Any member of $\mathfrak{D}$ that is the product of two nonzero members of $\mathfrak{D}$ neither of which is a unit is called **composite**. Any nonzero member of $\mathfrak{D}$ that is neither a unit nor composite is called **prime**. An integral domain $\mathfrak{D}$ is a **unique factorization domain** iff every nonzero nonunit member of $\mathfrak{D}$ can be expressed as a product of a finite number of prime members of $\mathfrak{D}$, and when so expressed is uniquely so expressed except for the order of the factors or multiplication of the factors by units.

In the real number system $\mathfrak{R}$ define the set Φ of all numbers of the form $a + b\sqrt{5}$, where $a, b \in \mathfrak{g}$. Then Φ is an integral domain. The following two facts are not difficult to prove (cf. [35], p. 144): (*i*) The units of Φ consist of all $a + b\sqrt{5}$ such that $| a^2 - 5b^2 | = 1$. (*ii*) If $a + b\sqrt{5}$ is a nonzero nonunit, then $| a^2 - 5b^2 | \geq 4$. Consequently, if $1 < | a^2 - 5b^2 | < 16, a + b\sqrt{5}$ is prime. In particular,

2, $1 + \sqrt{5}$, and $-1 + \sqrt{5}$ are all prime members of Φ since for each, $|a^2 - 5b^2| = 4$. Furthermore, the two factorings of 4,

$$2 \cdot 2 = (1 + \sqrt{5})(-1 + \sqrt{5}),$$

are distinct in the sense defined above: No factor of either member is a unit times either factor of the other member. (For details see [35], p. 145.)

9. Two numbers without a greatest common divisor.

In an integral domain $\mathfrak{D}$, a member m **divides** a member n, written $m \mid n$, iff there exists a member p of $\mathfrak{D}$ such that $mp = n$. A member d of $\mathfrak{D}$ is called a **greatest common divisor** of two members a and b of $\mathfrak{D}$ iff:

(i) $\qquad\qquad\qquad d \mid a \quad \text{and} \quad d \mid b;$

(ii) $\qquad\qquad\qquad c \mid a, c \mid b \Rightarrow c \mid d.$

In the integral domain Φ of the preceding example, the numbers 4 and $2(1 + \sqrt{5})$ have no greatest common divisor. (For details, see [35], pp. 145–146.)

10. A fraction that cannot be reduced to lowest terms uniquely.

If fractions are constructed from pairs of members of the integral domain Φ of Example 8, the fraction $2(1 + \sqrt{5})/4$ can be reduced to lowest terms in the following two ways:

$$\frac{2(1 + \sqrt{5})}{4} = \frac{1 + \sqrt{5}}{2} = \frac{2}{-1 + \sqrt{5}}.$$

The results are distinct in the sense that neither numerator is a unit times the other, and neither denominator is a unit times the other.

11. Functions continuous on a closed interval and failing to have familiar properties in case the number system is not complete.

We conclude this chapter with a collection of functions defined on a closed interval $[a, b] \subset \mathbb{Q}$ and having values in $\mathbb{Q}$. These examples would all be impossible if the rational number system $\mathbb{Q}$, which is not complete (cf. Example 3), were replaced by the real number system $\Re$, which is complete. The ordered field Q will be considered to be embedded in $\Re$ in order that symbols such as $\sqrt{2}$ can be used. The letter x is assumed to represent a rational number in every case.

a. *A function continuous on a closed interval and not bounded there (and therefore, since the interval is bounded, not uniformly continuous there).*

$$f(x) = \frac{1}{x^2 - 2}, \qquad 0 \leqq x \leqq 2.$$

b. *A function continuous and bounded on a closed interval but not uniformly continuous there.*

$$f(x) = \begin{cases} 0, & 0 \leqq x < \sqrt{2}, \\ 1, & \sqrt{2} < x \leqq 2. \end{cases}$$

c. *A function uniformly continuous (and therefore bounded) on a closed interval and not possessing a maximum value there.*

$$f(x) = x - x^3, \qquad 0 \leqq x \leqq 1.$$

d. *A function continuous on a closed interval and failing to have the intermediate value property.*

Example b; or $f(x) = x^2$ on [1, 2], which does not assume the value 2 intermediate between the values 1 and 4.

e. *A nonconstant differentiable function whose derivative vanishes identically over a closed interval.*

Example b.

f. *A differentiable function for which Rolle's theorem (and therefore the law of the mean) fails.*

Example c.

g. *A monotonic uniformly continuous nonconstant function having the intermediate value property, and whose derivative is identically 0 on an interval.*

This example is more difficult than the preceding ones. It can be constructed by means of the Cantor set defined and discussed in Chapter 8. For details, see Example 15, Chapter 8.

Chapter 2
Functions and Limits

Introduction

In this chapter it will be necessary to extend some of the definitions of Chapter 1, or to introduce new ones. Unless a specific statement to the contrary is made, all sets under consideration will be assumed to be subsets of $\mathfrak{R}$, the real number system, and all functions will be assumed to be real-valued functions of a real variable.

We start by extending unions and intersections to infinite collections of sets $A_1, A_2, \cdots$:

$$\bigcup_{n=1}^{+\infty} A_n = A_1 \cup A_2 \cup \cdots \equiv \{x \mid x \in A_n \text{ for at least one } n = 1, 2, \cdots\},$$

$$\bigcap_{n=1}^{+\infty} A_n = A_1 \cap A_2 \cap \cdots \equiv \{x \mid x \in A_n \text{ for every } n = 1, 2, \cdots\}.$$

A set A is **closed** iff it contains all its limit points; that is, iff there is no point of A' that is a limit point of A. A set A is **open** iff every point of A has a neighborhood lying entirely in A. A point p is a **frontier point** of a set A iff every neighborhood of p contains at least one point of A and at least one point of A'. The set of all frontier points of A is called the **frontier** of A, and is denoted $F(A)$. A point p is an **interior point** of a set A iff there exist a neighborhood of p that lies entirely in A. The set of interior points of A is called the **interior** of A, and is denoted $I(A)$. Any closed set A is the union of its interior and its frontier: $A = I(A) \cup F(A)$. The **closure** of A, denoted $\bar{A}$, is the union of the set A and the set of all limit

20

points of A. An **open covering** of a set A is a family $\{U_\alpha\}$ of open sets U_α whose union contains A; in this case $\{U_\alpha\}$ **covers** A. A set A is **compact** iff every open covering of A contains a finite subfamily that covers A. In the space $\mathfrak{R}$ a set is compact iff it is closed and bounded. (This is the Heine-Borel theorem and its converse; cf. [34], p. 202.)

A set A is **countable** iff A is finite or there exists a one-to-one correspondence whose domain is $\mathfrak{N}$, the set of natural numbers, and whose range is A .

An important property of the real number system is that, for any real number x, there exists a unique integer n such that

$$n \leqq x < n + 1, \quad \text{or} \quad x - 1 < n \leqq x.$$

Since n is determined uniquely as the greatest integer less than or equal to x, a function f is thereby defined, known as the **greatest integer function** or the **bracket function**, denoted $f(x) = [x]$, and equivalently defined as the integer $[x]$ satisfying

$$[x] \leqq x < [x] + 1, \quad \text{or} \quad x - 1 < [x] \leqq x.$$

Square brackets should be interpreted as indicating the bracket function *only* when an explicit statement to that effect is made.

A function f on $\mathfrak{R}$ into $\mathfrak{R}$ is **periodic with period** p iff $f(x + p) = f(x)$ for all $x \in \mathfrak{R}$. A function is **periodic** iff it is periodic with period p for some nonzero p.

Let a be a limit point of the domain D of a function f, and assume that $f(x)$ is bounded in some neighborhood of a, for $x \in D$. The **limit superior** and **limit inferior** of f at a, denoted $\overline{\lim}_{x \to a} f(x)$ and $\underline{\lim}_{x \to a} f(x)$, respectively, are defined in terms of the functions ϕ and ψ as follows: For $\delta > 0$,

$$\phi(\delta) \equiv \sup \{f(x) \mid x \in D \cap D(a, \delta)\},$$

$$\psi(\delta) \equiv \inf \{f(x) \mid x \in D \cap D(a, \delta)\},$$

$$\overline{\lim_{x \to a}} f(x) \equiv \lim_{\delta \to 0+} \phi(\delta) = \inf \{\phi(\delta) \mid \delta > 0\},$$

$$\underline{\lim_{x \to a}} f(x) \equiv \lim_{\delta \to 0+} \psi(\delta) = \sup \{\psi(\delta) \mid \delta > 0\}.$$

A function f is **upper semicontinuous** at a point $a \in D$ iff $\overline{\lim}_{x \to a} f(x)$ $\leq f(a)$; f is **lower semicontinuous** at a iff $\underline{\lim}_{x \to a} f(x) \geq f(a)$; f is **semicontinuous** at a iff f is either upper semicontinuous at a or lower semicontinuous at a.

A function f is **locally bounded at a point** a that is either a point or a limit point of the domain of f iff there exists a neighborhood of a on which f is bounded; f is **locally bounded** on a subset A of its domain iff f is locally bounded at every point of A.

Infinite limits $\pm \infty$, and limits of $f(x)$ as $x \to \pm \infty$, are defined as in the case of $\lim_{x \to a} f(x) = b$, except that (deleted) neighborhoods of infinity are used:

$$D(+\infty, N) \equiv (N, +\infty),$$

$$D(-\infty, N) \equiv (-\infty, N).$$

For example:

$$\lim_{x \to a} f(x) = +\infty \quad \text{iff} \quad \forall\, K\, \exists\, \delta > 0 \ni f(D \cap D(a, \delta)) \subset D(+\infty, K),$$

$$\lim_{x \to -\infty} f(x) = b \quad \text{iff} \quad \forall\, \varepsilon > 0\, \exists\, N \ni f(D \cap D(-\infty, N)) \subset N(b, \varepsilon).$$

Basic definitions of convergence and uniform convergence of infinite series, and the Weierstrass M-test for uniform convergence, will be assumed as known (cf. [34], pp. 381, 444, 445).

1. A nowhere continuous function whose absolute value is everywhere continuous.

$$f(x) \equiv \begin{cases} 1 & \text{if } x \text{ is rational,} \\ -1 & \text{if } x \text{ is irrational.} \end{cases}$$

2. A function continuous at one point only. (Cf. Example 22.)

$$f(x) \equiv \begin{cases} x & \text{if } x \text{ is rational,} \\ -x & \text{if } x \text{ is irrational.} \end{cases}$$

The only point of continuity is 0.

3. For an arbitrary noncompact set, a continuous and unbounded function having the set as domain.

(a) If A is an unbounded set of real numbers, let

$$f(x) \equiv x, \quad x \in A.$$

(b) If A is a bounded set of real numbers that is not closed, let c be a limit point of A but not a member of A, and let

$$f(x) \equiv \frac{1}{x - c}, \quad x \in A.$$

If f is continuous on a *compact* set A, then f is bounded there (cf. [36], p. 80).

4. For an arbitrary noncompact set, an unbounded and locally bounded function having the set as domain.

Example 3.

If f is locally bounded on a compact set A, then f is bounded there.

5. A function that is everywhere finite and everywhere locally unbounded.

If x is a rational number equal to m/n, where m and n are integers such that the fraction m/n is in lowest terms and $n > 0$, then m and n are uniquely determined (cf. [35], p. 53). Therefore the following function is well defined:

$$f(x) = \begin{cases} n & \text{if } x \text{ is rational}, \quad x = m/n \text{ in lowest terms}, \quad n > 0; \\ 0 & \text{if } x \text{ is irrational.} \end{cases}$$

If f were bounded in $N(a, \varepsilon)$, then for all m/n in $N(a, \varepsilon)$ the denominators n would be bounded, and hence the numerators m would be too. But this would permit only finitely many rational numbers in the interval $N(a, \varepsilon)$. (Contradiction.) (Cf. Example 27, Chapter 8, for a function incorporating these and more violent pathologies. Also cf. Example 29, Chapter 8.)

6. For an arbitrary noncompact set, a continuous and bounded function having the set as domain and assuming no extreme values.

(a) If A is an unbounded set of real numbers, let

$$f(x) \equiv \frac{x^2}{x^2 + 1}, \qquad x \in A.$$

Then $f(x)$ has no maximum value on A. If $f(x)$ is defined

$$f(x) \equiv (-1)^{[|x|]} \frac{x^2}{x^2 + 1}, \qquad x \in A,$$

where $[|x|]$ is the greatest integer less than or equal to $|x|$, then $f(x)$ has neither a maximum value nor a minimum value on A.

(b) If A is a bounded set of real numbers that is not closed, let c be a limit point of A but not a member of A, and let

$$f(x) \equiv -|x - c|, \qquad x \in A.$$

Then $f(x)$ has no maximum value on A. If $f(x)$ is defined

$$f(x) = (-1)^{[1/|x-c|]} \{ L - |x - c| \},$$

where brackets are once again used to represent the "bracket function," and L is the length of some interval containing A, then $f(x)$ has neither a maximum nor a minimum value on A.

7. A bounded function having no relative extrema on a compact domain.

Let the compact domain be the closed interval $[0, 1]$, and for $x \in [0, 1]$, define

$$f(x) \equiv \begin{cases} \dfrac{(-1)^n n}{n + 1} & \text{if } x \text{ is rational, } x = m/n \text{ in lowest terms, } n > 0. \\ \\ 0 & \text{if } x \text{ is irrational}. \end{cases}$$

Then in every neighborhood of every point of $[0, 1]$ the values of f come arbitrarily close to the numbers 1 and -1 while always lying between them. (Cf. [14], p. 127.)

8. A bounded function that is nowhere semicontinuous.

The function of Example 7 is nowhere upper semicontinuous since $\overline{\lim}_{x \to a} f(x)$ is everywhere equal to 1 and therefore nowhere $\leq f(a)$. Similarly, this function is nowhere lower semicontinuous. (Notice that the function of Example 1 is upper semicontinuous at a iff a is rational and lower semicontinuous at a iff a is irrational.)

9. A nonconstant periodic function without a smallest positive period.

The periods of the function of Example 1 are the rational numbers.

The periods of any real-valued function with domain ℜ form an additive group (that is, the set of periods is closed with respect to subtraction). This group is either dense (as in the present example) or discrete, consisting of all integral multiples of a least positive member. This latter case always obtains for a nonconstant periodic function with domain ℜ that has at least *one* point of continuity. (Cf. [36], p. 549.)

10. An irrational function.

The function $\sqrt{x}$ is not a rational function (cf. Example 4, Chapter 1) since it is undefined for $x < 0$.

The function $[x]$ is not a rational function since it has discontinuities at certain points where it is defined.

The function $|x|$ is not a rational function since it fails to have a derivative at a point at which it is defined.

The function $\sqrt{x^2 + 1}$ is not a rational function. This can be seen as follows: If $\sqrt{x^2 + 1} = f(x)/g(x)$ for all x, then $\sqrt{x^2 + 1}/x = f(x)/xg(x)$ for all $x \neq 0$, and hence $\lim_{x \to +\infty} f(x)/xg(x) = 1$. This means that $f(x)$ and $xg(x)$ are polynomials of the same degree, and therefore $\lim_{x \to -\infty} f(x)/xg(x) = 1$, whereas $\lim_{x \to -\infty} \sqrt{x^2 + 1}/x = -1$. (Contradiction.)

11. A transcendental function.

A function f is **algebraic** iff ∃ a polynomial $p(u) = \sum_{k=0}^{n} a_k(x)u^k$, whose coefficients $a_0(x)$, $a_1(x)$, $\cdots$, $a_n(x)$ are real polynomials (that is, their coefficients are all real) not all of which are identically zero and such that the composite function $p(f(x))$ vanishes identically on the domain of f. A function is **transcendental** iff it is not algebraic.

An example of a transcendental function is e^x, for if it is assumed that

$$a_0(x) + a_1(x)e^x + \cdots + a_n(x)e^{nx},$$

where $a_0(x)$ is not the zero polynomial, vanishes identically, a contradiction is got by taking the limit as $x \to -\infty$ and using l'Hospital's rule on indeterminate forms to infer the impossible conclusion

$$\lim_{x \to -\infty} a_0(x) = 0.$$

25

Another example is sin x, since if

$$b_0(x) + b_1(x) \sin x + \cdots + b_n(x) \sin^n x,$$

where $b_0(x)$ is not the zero polynomial, vanishes identically, then $b_0(k\pi) = 0$ for all integral k. (Contradiction.)

Other examples (for similar reasons) are $\ln x$ (the inverse of e^x) and the remaining trigonometric functions.

The following functions listed under Example 10 as irrational are algebraic: $\sqrt{x}$, $|x|$ $(|x| = \sqrt{x^2})$, and $\sqrt{x^2 + 1}$.

12. Functions $y = f(u)$, $u \in \mathfrak{R}$, and $u = g(x)$, $x \in \mathfrak{R}$, whose composite function $y = f(g(x))$ is everywhere continuous, and such that

$$\lim_{u \to b} f(u) = c, \ \lim_{x \to a} g(x) = b, \ \lim_{x \to a} f(g(x)) \neq c.$$

If

$$f(u) \equiv \begin{cases} 0 & \text{if} \quad u \neq 0, \quad u \in \mathfrak{R}, \\ 1 & \text{if} \quad u = 0, \end{cases}$$

then $\lim_{u \to 0} f(u) = 0$. If $g(x) \equiv 0$ for all $x \in \mathfrak{R}$, then, $f(g(x)) = 1$ for all x, and hence $\lim_{x \to 0} f(g(x)) = 1$.

This counterexample becomes impossible in case the following condition is added: $x \neq a \Rightarrow g(x) \neq b$.

13. Two uniformly continuous functions whose product is not uniformly continuous.

The functions x and sin x are uniformly continuous on $\mathfrak{R}$ since their derivatives are bounded, but their product x sin x is not uniformly continuous on $\mathfrak{R}$.

In case *both* functions f and g are bounded on a common domain D and uniformly continuous on D, their product fg is also uniformly continuous on D. Since any function uniformly continuous on a bounded set is bounded there, it follows that the present counterexample is possible only when the common domain under consideration is unbounded and at least one of the functions is unbounded.

14. A function continuous and one-to-one on an interval and whose inverse is not continuous.

For this example it is necessary that the interval *not* be a closed bounded interval (cf. [34], p. 192), and that the function *not* be strictly real-valued (cf. [34], p. 50, p. 52, Ex. 25). Our example in this case is a complex-valued function $z = f(x)$ of the real variable x, with continuity defined exactly as in the case of a real-valued function of a real variable, where the absolute value of the complex number $z = (a, b)$ is defined

$$| z | = |(a, b)| \equiv (a^2 + b^2)^{1/2}.$$

Let the function $z = f(x)$ be defined:

$$z = f(x) \equiv (\cos x, \sin x), \qquad 0 \leqq x < 2\pi.$$

Then f maps the half-open interval $[0, 2\pi)$ onto the unit circle $| z | = 1$ continuously and in a one-to-one manner. Since the unit circle is compact the inverse mapping cannot be continuous (cf. [34], p. 192), and fails to be continuous at the point $(1, 0)$.

15. A function continuous at every irrational point and discontinuous at every rational point.

If x is a rational number equal to m/n, where m and n are integers such that the fraction m/n is in lowest terms and $n > 0$, let $f(x)$ be defined to be equal to $1/n$. Otherwise, if x is irrational, let $f(x) \equiv 0$. (Cf. [34], p. 124.)

It will be shown in Example 10, Chapter 8, that there does not exist a function continuous at every rational point and discontinuous at every irrational point.

16. A semicontinuous function with a dense set of points of discontinuity.

The function of Example 15 is upper semicontinuous at every point a, since

$$\overline{\lim_{x \to a}} f(x) = 0 \leqq f(a).$$

17. A function with a dense set of points of discontinuity every one of which is removable.

If a is a rational number and if the function of Example 15 is redefined at a to have the value zero, then, since

$$\lim_{x \to a} f(x) = 0 = f(a),$$

as redefined, f becomes continuous at a.

18. A monotonic function whose points of discontinuity form an arbitrary countable (possibly dense) set.

If A is an arbitrary nonempty countable set of real numbers, a_1, a_2, a_3, $\cdots$, let $\sum p_n$ be a finite or convergent infinite series of positive numbers with sum p (the series being finite iff A is finite, and having as many terms as A has members). If A is bounded below and $x <$ every point of A, let $f(x) \equiv 0$. Otherwise, define $f(x)$ to be the sum of all terms p_m of $\sum p_n$ such that $a_m \leq x$. Then f is increasing on $\Re$, continuous at every point not in A, and discontinuous with a jump equal to p_n at each point a_n (that is, $\lim_{x \to a_n+} f(x) - \lim_{x \to a_n-} f(x) = p_n$).

It should be noted that for monotonic functions this example illustrates the most that can be attained by way of discontinuities: for any monotonic function the set of points of discontinuities is countable (cf. [36], p. 59, Ex. 29). Example 1 shows that without monotonicity the set of points of discontinuity may be the entire domain.

19. A function with a dense set of points of continuity, and a dense set of points of discontinuity no one of which is removable.

In Example 18, let the set A be the set $\mathbb{Q}$ of all rational numbers.

20. A one-to-one correspondence between two intervals that is nowhere monotonic.

Let $f(x)$ be defined for $0 \leq x \leq 1$:

$$f(x) \equiv \begin{cases} x & \text{if } x \text{ is rational,} \\ 1 - x & \text{if } x \text{ is irrational.} \end{cases}$$

Then there is no subinterval of $[0, 1]$ on which f is monotonic. The range of f is again the interval $[0, 1]$, and f is one-to-one.

A function having these properties and mapping the interval $[a, b]$

onto the interval $[c, d]$ is

$$g(x) \equiv \begin{cases} c + (d - c)\dfrac{x - a}{b - a} & \text{if } \dfrac{x - a}{b - a} \text{ is rational,} \\[2ex] d + (c - d)\dfrac{x - a}{b - a} & \text{if } \dfrac{x - a}{b - a} \text{ is irrational.} \end{cases}$$

21. A continuous function that is nowhere monotonic.

Let $f_1(x) \equiv |x|$ for $|x| \leq \frac{1}{2}$, and let $f_1(x)$ be defined for other values of x by periodic continuation with period 1, i.e., $f_1(x + n) = f_1(x)$ for every real number x and integer n. For $n > 1$ define $f_n(x) \equiv 4^{-n+1}f_1(4^{n-1}x)$, so that for every positive integer n, f_n is a periodic function of period 4^{-n+1}, and maximum value $\frac{1}{2} \cdot 4^{-n+1}$. Finally, define f with domain $\Re$:

$$f(x) \equiv \sum_{n=1}^{+\infty} f_n(x) = \sum_{n=1}^{+\infty} \frac{f_1(4^{n-1}x)}{4^{n-1}}.$$

Since $|f_n(x)| \leq \frac{1}{2} \cdot 4^{-n+1}$, by the Weierstrass M-test this series converges uniformly on $\Re$, and f is everywhere continuous. For any point a of the form $a = k \cdot 4^{-m}$, where k is an integer and m is a positive integer, $f_n(a) = 0$ for $n > m$, and hence $f(a) = f_1(a) + \cdots + f_m(a)$. For any positive integer m, let h_m be the positive number 4^{-2m-1}. Then $f_n(a + h_m) = 0$ for $n > 2m + 1$, and hence

$$f(a + h_m) - f(a) = [f_1(a + h_m) - f_1(a)] + \cdots$$
$$+ [f_m(a + h_m) - f_m(a)]$$
$$+ f_{m+1}(a + h_m) + \cdots + f_{2m+1}(a + h_m)$$
$$\geq -mh_m + (m + 1)h_m = h_m > 0.$$

Similarly,

$$f(a - h_m) - f(a) \geq -mh_m + (m + 1)h_m = h_m > 0.$$

Since members of the form $a = k \cdot 4^{-m}$ are dense, it follows that in no open interval is f monotonic.

The above typifies constructions involving the condensation of singularities.

22. A function whose points of discontinuity form an arbitrary given closed set.

If A is a closed set, define the set B:

$$x \in B \text{ iff } \begin{cases} x \in F(A) \quad \text{or} \\ x \in I(A) \cap \mathbb{Q}, \end{cases}$$

and define the function f:

$$f(x) \equiv \begin{cases} 1 & \text{if } x \in B, \\ 0 & \text{if } x \notin B. \end{cases}$$

If $c \in A$, f is discontinuous at c: if $c \in F(A)$ then $f(c) = 1$ while x is a limit point of the set $\mathbb{R} \setminus A$, on which f is identically 0; if $c \in I(A) \cap \mathbb{Q}$ then $f(x) = 1$ while c is a limit point of the set $I(A) \setminus \mathbb{Q}$, on which f is identically 0; if $c \in I(A) \setminus \mathbb{Q}$, then $f(c) = 0$ while c is a limit point of the set $I(A) \cap \mathbb{Q}$, on which f is identically 1. The function f is continuous on the set $\mathbb{R} \setminus A$, since this set is open and f is a constant there.

23. A function whose points of discontinuity form an arbitrary given F_σ set. (Cf. Example 8, Chapter 4, and Examples 8, 10, and 22, Chapter 8.)

A set A is said to be an F_σ set iff it is a countable union of closed sets (cf. Example 8, Chapter 8). For a given F_σ set $A = A_1 \cup A_2 \cup \cdots$, where $A_1, A_2, \cdots$ are closed and $A_n \subset A_{n+1}$ for $n = 1, 2, \cdots$, let A_0 denote the empty set $\emptyset$, and define the disjoint sets B_n, $n = 1, 2, \cdots$:

$$x \in B_n \text{ iff } \begin{cases} x \in (A_n \setminus A_{n-1}) \setminus I(A_n \setminus A_{n-1}) \quad \text{or} \\ x \in I(A_n \setminus A_{n-1}) \cap \mathbb{Q}. \end{cases}$$

Let the function f be defined:

$$f(x) \equiv \begin{cases} 2^{-n} & \text{if } x \in B_n, \\ 0 & \text{if } x \notin B_1 \cup B_2 \cup \cdots. \end{cases}$$

If $c \in A$, f is discontinuous at c: if $c \in (A_n \setminus A_{n-1}) \setminus I(A_n \setminus A_{n-1})$ then $f(c) = 2^{-n}$ while c is a limit point of a set on which f has values differing from 2^{-n} by at least 2^{-n-1}; if $c \in I(A_n \setminus A_{n-1}) \cap \mathbb{Q}$ then $f(c) = 2^{-n}$ while c is a limit point of the set $I(A_n \setminus A_{n-1}) \setminus \mathbb{Q}$ on which f is identically 0; if $c \in I(A_n \setminus A_{n-1}) \setminus Q$, then $f(c) = 0$ while c is a limit point of the set $I(A_n \setminus A_{n-1}) \cap Q$ on which f is identically 2^{-n}. If $c \notin A$, f is continuous at c: if $\varepsilon > 0$ is given, choose N such that $2^{-N} < \varepsilon$; then choose a neighborhood of c that excludes $A_1, A_2, \cdots, A_N$ and inside which, therefore, $f(x) < 2^{-N} < \varepsilon$.

It should be noted that for *any* function f on $\mathfrak{R}$ into $\mathfrak{R}$ the set of points of discontinuity is an F_σ set (cf. [36], p. 84, Exs. 30–33, p. 332, Ex. 41).

24. A function that is not the limit of any sequence of continuous functions. (Cf. Example 10, Chapter 4.)

The function f in Example 1 has the property that there is no sequence $\{f_n\}$ of continuous functions such that $\lim_{n\to+\infty} f_n(x) = f(x)$ for all real x, but the proof is not elementary. For a discussion and references, see [10], pp. 99–102. The idea is that f is everywhere discontinuous, while any function that is the limit of a sequence of continuous functions must have a dense set of points of continuity.

The characteristic function of the set Q of rational numbers is the limit of a sequence $\{g_n\}$ of functions each of which is the limit of a sequence $\{h_n\}$ of continuous functions, as follows: If $\{r_n\}$ is a sequence that is a one-to-one correspondence with domain $\mathfrak{N}$ and range Q, define

$$g_n(x) \equiv \begin{cases} 1 & \text{if } x = r_1, r_2, \cdots, \text{ or } r_n, \\ 0 & \text{otherwise.} \end{cases}$$

Each function g_n is the limit of a sequence of continuous functions each of which is equal to 1 where $g_n(x) = 1$, equal to 0 on closed subintervals interior to the intervals between consecutive points where $g_n(x) = 1$, and linear between consecutive points that are either endpoints of such closed subintervals or points where $g_n(x) = 1$. Notice that for each x, $\{g_n(x)\}$ is increasing, while the sequence that converges to $g_n(x)$ can be chosen to be decreasing.

25. A function with domain [0, 1] whose range for every nondegenerate subinterval of [0, 1] is [0, 1]. (Cf. Example 27, Chapter 8.)

A function having this property was constructed by H. Lebesgue (cf. [28], p. 90) and is described in [10], p. 71. (Also cf. [14], p. 228.)

I. Functions of a Real Variable

If x is an arbitrary number in $[0, 1]$, let its decimal expansion be

$$x = 0.a_1a_2a_3 \cdots$$

where, in case x can be expressed ambiguously by either a terminating decimal or one with indefinitely repeated 9's, it is immaterial which expansion is chosen. For definiteness suppose the terminating expansion is consistently chosen. The value $f(x)$ depends on whether or not the decimal $0.a_1a_3a_5 \cdots$ is repeating or not — that is, on whether the number $0.a_1a_3a_5 \cdots$ is rational or not (cf. [35], p. 178):

$$f(x) \equiv \begin{cases} 0 & \text{if} \quad 0.a_1a_3a_5 \cdots \text{ is irrational,} \\ 0.a_{2n}a_{2n+2}a_{2n+4} \cdots & \text{if } 0.a_1a_3a_5 \cdots \text{ is rational with its} \\ & \text{first repeating segment beginning} \\ & \text{with } a_{2n-1}. \end{cases}$$

Let I be an arbitrary subinterval of $[0, 1]$, and choose the digits $a_1, a_2, \cdots, a_{2n-2}$ so that both $0.a_1a_2 \cdots a_{2n-2}0$ and $0.a_1a_2 \cdots a_{2n-2}1$ belong to I and such that a_{2n-3} is different from both 0 and 1. If $y = 0.b_1b_2b_3 \cdots$ is an arbitrary point in $[0, 1]$, we have only to define $a_{2n-1} = a_{2n+1} = \cdots = a_{4n-5} = 0$ and $a_{4n-3} = 1$, with subsequent a's with odd subscripts defined by cyclic repetition in groups of n, to obtain a number

$$x = 0.a_1a_2a_3 \cdots a_{2n-1}b_1a_{2n+1}b_2a_{2n+3} \cdots$$

belonging to the interval I and such that the expansion

$$0.a_1a_3a_5 \cdots a_{2n-3}a_{2n-1}a_{2n+1} \cdots$$

is a periodic decimal whose first period starts with a_{2n-1}, and consequently such that

$$f(x) = 0.b_1b_2b_3 \cdots .$$

The graph of f is dense in the unit square $[0, 1] \times [0, 1]$, although each vertical segment $\{x\} \times [0, 1]$ meets the graph in exactly one point.

A function whose range on every nonempty open interval is $\mathfrak{R}$ and that is equal to zero almost everywhere (and hence is measurable) is given in Example 27, Chapter 8. (Also cf. Example 26, below.)

Since the unit interval $[0, 1]$ contains infinitely many disjoint open intervals (no two have a point in common — for example, $\left(\dfrac{1}{n+1}, \dfrac{1}{n}\right)$, $n = 1, 2, \cdots$) — the function f of the present example takes on every one of its values infinitely many times. Another example of a function that assumes every value infinitely many times is given in Example 9, Chapter 10.

26. A discontinuous linear function.

A function f on $\Re$ into $\Re$ is said to be **linear** iff $f(x + y) = f(x) + f(y)$ for all $x, y \in \Re$. A function that is linear and *not* continuous must be very discontinuous indeed. In fact, its graph must be dense in the plane $\Re \times \Re$. For a discussion of this phenomenon, and further references, see [10], pp. 108–113. In case f is continuous it must have the form $f(x) = cx$, as can be shown by considering in succession the following classes of numbers: $\mathfrak{N}, \mathcal{I}, \mathcal{Q}, \Re$.

Construction of a discontinuous linear function can be achieved by use of a Hamel basis for the linear space of the real numbers $\Re$ over the rational numbers $\mathcal{Q}$ (cf. references 29, 30, and 32 of [10]). The idea is that this process provides a set $S = \{r_\alpha\}$ of real numbers r_α such that every real number x is a *unique* linear combination of a finite number of members of S with rational coefficients p_α: $x = p_{\alpha_1} r_{\alpha_1} + \cdots + p_{\alpha_k} r_{\alpha_k}$. The function f can now be defined:

$$f(x) \equiv p_{\alpha_1} + \cdots + p_{\alpha_k},$$

since the representation of x as a linear combination is unique. The linearity of f follows directly from the definition, and the fact that f is not continuous follows from the fact that its values are all rational but not all equal (f fails to have the intermediate value property).

27. For each $n \in \mathfrak{N}$, $n(2n + 1)$ functions $\phi_{ij}(x_j)$, $j = 1, 2,$ $\cdots, n, i = 1, 2, \cdots, 2n + 1$, satisfying:

(a) All $\phi_{ij}(x_j)$ are continuous on $[0, 1]$.

(b) For any function $f(x_1, x_2, \cdots, x_n)$ continuous for $0 \leqq x_1, x_2, \cdots, x_n \leqq 1$, there are $2n + 1$ functions ψ_i, $i = 1, 2, \cdots, 2n + 1$, each continuous on $\Re$, such that

$$f(x_1, x_2, \cdots, x_n) = \sum_{i=1}^{2n+1} \psi_i \left(\sum_{j=1}^{n} \phi_{ij}(x_j) \right).$$

I. Functions of a Real Variable

This theorem is due to A. N. Kolmogorov [26] and resolves a famous problem posed by D. Hilbert. Stated as a solution of Hilbert's (thirteenth) problem, the above result reads: *Every continuous function $f(x_1, x_2, \cdots, x_n)$ of n real variables, $0 \leqq x_1, x_2, \cdots, x_n \leqq 1$, may be expressed as a sum* (the sum $\sum_{i=1}^{2n+1}$ above) *of the composites of continuous functions of single variables and sums of continuous functions of single variables* (the sums $\sum_{j=1}^{n}$ above).

The proof is highly ingenious, although it is accessible to any reader with the patience to trace through a rather straightforward multiple induction.

We note only that the functions ϕ_{ij} are universal in that they do not depend on f. The functions ψ_i, while not independent of f, are not uniquely determined by f (even after the functions ϕ_{ij} have been constructed). Details will be found in the cited reference.

Chapter 3
Differentiation

Introduction

In some of the examples of this chapter the word *derivative* is permitted to be applied to the infinite limits

$$\lim_{h \to 0} \frac{f(x + h) - f(x)}{h} = +\infty, \lim_{h \to 0} \frac{f(x + h) - f(x)}{h} = -\infty.$$

However, the term *differentiable function* is used only in the strict sense of a function having a finite derivative at each point of its domain. A function is said to be **infinitely differentiable** iff it has (finite) derivatives of all orders at every point of its domain.

The exponential function with base e is alternatively denoted e^x and exp (x).

As in Chapter 2, all sets, including domains and ranges, will be assumed to be subsets of $\Re$ unless explicit statement to the contrary is made. This assumption will remain valid through Part I of this book, that is, through Chapter 8.

1. A function that is not a derivative.

The signum function (cf. the Introduction, Chapter 1) or, indeed, any function with jump discontinuities, has no primitive — that is, fails to be the derivative of any function — since it fails to have the intermediate value property enjoyed by continuous functions and derivatives alike (cf. [34], p. 84, Ex. 40). An example of a discontinuous derivative is given next.

2. A differentiable function with a discontinuous derivative.

The function

$$f(x) \equiv \begin{cases} x^2 \sin \dfrac{1}{x} & \text{if } x \neq 0, \\ 0 & \text{if } x = 0, \end{cases}$$

has as its derivative the function

$$f'(x) = \begin{cases} 2x \sin \dfrac{1}{x} - \cos \dfrac{1}{x} & \text{if } x \neq 0, \\ 0 & \text{if } x = 0, \end{cases}$$

which is discontinuous at the origin.

3. A discontinuous function having everywhere a derivative (not necessarily finite).

For such an example to exist the definition of derivative must be extended to include the limits $\pm \infty$. If this is done, the discontinuous signum function (Example 1) has the derivative

$$g(x) = \begin{cases} 0 & \text{if } x \neq 0, \\ +\infty & \text{if } x = 0. \end{cases}$$

4. A differentiable function having an extreme value at a point where the derivative does not make a simple change in sign.

The function

$$f(x) \equiv \begin{cases} x^4 \left(2 + \sin \dfrac{1}{x} \right) & \text{if } x \neq 0, \\ 0 & \text{if } x = 0 \end{cases}$$

has an absolute minimum value at $x = 0$. Its derivative is

$$f'(x) \equiv \begin{cases} x^2 \left[4x \left(2 + \sin \dfrac{1}{x} \right) - \cos \dfrac{1}{x} \right] & \text{if } x \neq 0, \\ 0 & \text{if } x = 0, \end{cases}$$

which has both positive and negative values in every neighborhood of the origin. In no interval of the form $(a, 0)$ or $(0, b)$ is f monotonic.

5. A differentiable function whose derivative is positive at a point but which is not monotonic in any neighborhood of the point.

The function

$$f(x) \equiv \begin{cases} x + 2x^2 \sin \dfrac{1}{x} & \text{if } x \neq 0, \\ 0 & \text{if } x = 0 \end{cases}$$

has the derivative

$$f'(x) = \begin{cases} 1 + 4x \sin \dfrac{1}{x} - 2 \cos \dfrac{1}{x} & \text{if } x \neq 0, \\ 1 & \text{if } x = 0. \end{cases}$$

In every neighborhood of 0 the function $f'(x)$ has both positive and negative values.

6. A function whose derivative is finite but unbounded on a closed interval.

The function

$$f(x) \equiv \begin{cases} x^2 \sin \dfrac{1}{x^2} & \text{if } x \neq 0, \\ 0 & \text{if } x = 0 \end{cases}$$

has the derivative

$$f'(x) = \begin{cases} 2x \sin \dfrac{1}{x^2} - \dfrac{2}{x} \cos \dfrac{1}{x^2} & \text{if } x \neq 0, \\ 0 & \text{if } x = 0, \end{cases}$$

which is unbounded on $[-1, 1]$.

7. A function whose derivative exists and is bounded but possesses no (absolute) extreme values on a closed interval.

The function

$$f(x) \equiv \begin{cases} x^4 e^{-\frac{1}{2}x^2} \sin \dfrac{8}{x^3} & \text{if } x \neq 0, \\ 0 & \text{if } x = 0 \end{cases}$$

has the derivative

$$f'(x) = \begin{cases} e^{-\frac{1}{4}x^2}\left[(4x^3 - \tfrac{1}{2}x^5)\sin\dfrac{8}{x^3} - 24\cos\dfrac{8}{x^3} \right] & \text{if } x \neq 0, \\ 0 & \text{if } x = 0. \end{cases}$$

In every neighborhood of the origin this derivative has values arbitrarily near both 24 and -24. On the other hand, for $0 < h \equiv |x| \leqq 1$ (cf. [34], p. 83, Ex. 29),

$$0 < e^{-\frac{1}{4}x^2} < 1 - \frac{1}{4}h^2 e^{-\frac{1}{4}h^2} < 1 - \frac{3}{16}h^2,$$

and

$$\left| \left(4x^3 - \frac{1}{2}x^5\right)\sin\frac{8}{x^3} - 24\cos\frac{8}{x^3} \right| \leqq 24 + \frac{9}{2}h^3.$$

Therefore $0 < h \leqq 1$ implies

$$|f'(x)| < \left(1 - \frac{3}{16}h^2\right)\left(24 + \frac{9}{2}h^3\right) < 24 - \frac{9}{2}h^2(1 - h) \leqq 24.$$

Therefore, on the closed interval $[-1, 1]$ the range of the function f' has supremum equal to 24 and infimum equal to -24, and neither of these numbers is assumed as a value of f'.

8. A function that is everywhere continuous and nowhere differentiable.

The function $|x|$ is everywhere continuous but it is not differentiable at $x = 0$. By means of translates of this function it is possible to define everywhere continuous functions that fail to be differentiable at each point of an arbitrarily given finite set. In the following paragraph we shall discuss an example using an infinite set of translates of the function $|x|$.

The function of Example 21, Chapter 2, is nowhere differentiable. To see this let a be an arbitrary real number, and for any positive integer n, choose h_n to be either 4^{-n-1} or -4^{-n-1} so that $|f_n(a + h_n) - f_n(a)| = |h_n|$. Then $|f_m(a + h_n) - f_m(a)|$ has this same value $|h_n|$ for all $m \leqq n$, and vanishes for $m > n$. Hence the difference quotient $(f(a + h_n) - f(a))/h_n$ is an integer that is even if n is even and odd if n is odd. It follows that

$$\lim_{n \to +\infty} \frac{f(a + h_n) - f(a)}{h_n}$$

cannot exist, and therefore that $f'(a)$ cannot exist as a finite limit.

The first example of a continuous nondifferentiable function was given by K. W. T. Weierstrass (German, 1815–1897):

$$f(x) = \sum_{n=0}^{+\infty} b^n \cos (a^n \pi x),$$

where b is an odd integer and a is such that $0 < a < 1$ and $ab > 1 + \frac{3}{2}\pi$. The example presented above is a modification of one given in 1930 by B. L. Van der Waerden (cf. [48], p. 353). There are now known to be examples of continuous functions that have nowhere a one-sided finite or infinite derivative. For further discussion of these examples, and references, see [48], pp. 350–354, [10], pp. 61–62, 115, 126, and [21], vol. II, pp. 401–412.

The present example, as described in Example 21, Chapter 2, was shown to be nowhere monotonic. For an example of a function that is everywhere differentiable and nowhere monotonic, see [21], vol. II, pp. 412–421. Indeed, this last example gives a very elaborate construction of a function that is everywhere differentiable and has a dense set of relative maxima and a dense set of relative minima.*

9. A differentiable function for which the law of the mean fails.

Again, we must go beyond the real number system for the range of such a function. The complex-valued function of a real variable x,

$$f(x) \equiv \cos x + i \sin x,$$

is everywhere continuous and differentiable (cf. [34], pp. 509–513), but there exist no a, b, and ξ such that $a < \xi < b$ and

$$(\cos b + i \sin b) - (\cos a + i \sin a) = (-\sin \xi + i \cos \xi)(b - a).$$

Assuming that the preceding equation *is* possible, we equate the squares of the moduli (absolute values) of the two members:

---

*See also A. Denjoy, *Bull. Soc. Math. France*, 43 (1915), pp. 161–248 (228ff.).

$$(\cos b - \cos a)^2 + (\sin b - \sin a)^2 = (b - a)^2$$

or, with the aid of elementary identities:

$$\sin^2 \frac{b - a}{2} = \left(\frac{b - a}{2}\right)^2.$$

Since there is no positive number h such that $\sin h = h$ (cf. [34], p. 78), a contradiction has been obtained.

10. An infinitely differentiable function of x that is positive for positive x and vanishes for negative x.
The function

$$f(x) \equiv \begin{cases} e^{-1/x^2} & \text{if } x > 0, \\ 0 & \text{if } x \leq 0 \end{cases}$$

is infinitely differentiable, all of its derivatives at $x = 0$ being equal to 0 (cf. [34], p. 108, Ex. 52).

11. An infinitely differentiable function that is positive in the unit interval and vanishes outside.

$$f(x) \equiv \begin{cases} e^{-1/x^2(1-x)^2} & \text{if } 0 < x < 1, \\ 0 & \text{otherwise.} \end{cases}$$

12. An infinitely differentiable "bridging function," equal to 1 on $[1, +\infty)$, equal to 0 on $(-\infty, 0]$, and strictly monotonic on $[0, 1]$.

$$f(x) = \begin{cases} \exp\left[-\dfrac{1}{x^2} \exp\left(-\dfrac{1}{(1 - x)^2}\right)\right] & \text{if } 0 < x < 1, \\ 0 & \text{if } x \leq 0, \\ 1 & \text{if } x \geq 1. \end{cases}$$

13. An infinitely differentiable monotonic function f such that

$$\lim_{x \to +\infty} f(x) = 0, \qquad \lim_{x \to +\infty} f'(x) \neq 0.$$

If the word *monotonic* is deleted there are trivial examples, for instance $(\sin x^2)/x$. For a monotonic example, let $f(x)$ be defined to be equal to 1 for $x \leqq 1$, equal to $1/n$ on the closed interval $[2n - 1, 2n]$, for $n = 1, 2, \cdots$, and on the intervening intervals $(2n, 2n + 1)$ define $f(x)$ by translations of the bridging function of Example 12, with appropriate negative factors for changes in the vertical scale.

Chapter 4
Riemann Integration

Introduction

The definition of Riemann-integrability and the Riemann (or definite) integral of a function f defined on a closed interval $[a, b]$, together with the principal elementary properties of this integral, will be assumed known. The same is the case for the standard improper integrals and, in Example 14, for the Riemann-Stieltjes integral.

In some of the examples of this chapter the concept of measure zero is important. A set $A \subset \mathfrak{R}$ is said to be of **measure zero** iff for any $\varepsilon > 0$ there is an open covering of A consisting of a countable collection of open intervals whose lengths form a convergent infinite series with sum less than ε. The interior of every set of measure zero is empty. A point-property is said to hold **almost everywhere** iff the set where the property *fails* is of measure zero. A function f whose domain is a closed interval $[a, b]$ is Riemann-integrable there iff it is bounded and continuous almost everywhere (cf. [36], p. 153, Ex. 54).

1. A function defined and bounded on a closed interval but not Riemann-integrable there.

The characteristic function of the set $\mathbb{Q}$ of rational numbers, restricted to the closed interval $[0, 1]$, is not Riemann-integrable there (cf. [34], p. 112).

2. A Riemann-integrable function without a primitive.

The signum function (Example 1, Chapter 3) restricted to the interval $[-1, 1]$ is integrable there, but has no primitive there.

3. A Riemann-integrable function without a primitive on any interval.

Example 18, Chapter 2, with $A = \mathbb{Q} \cap [0, 1]$, is integrable on [0, 1] since it is monotonic there, but has no primitive on any sub-interval of [0, 1] since its points of jump discontinuity are dense there.

4. A function possessing a primitive on a closed interval but failing to be Riemann-integrable there. (Cf. Example 35, Chapter 8.)

The function f of Example 6, Chapter 3, is an example of a function having a (finite) derivative $g(x)$ at each point x of a closed interval I. The function g, therefore, has a primitive but since g is unbounded it is not Riemann-integrable on I.

The two preceding examples (Examples 3 and 4) are of interest in connection with the Fundamental Theorem of Calculus. One form of this theorem states that if a function $f(x)$ (*i*) *is integrable* on the interval $[a, b]$ and (*ii*) *has a primitive* $F(x)$ there ($F'(x) = f(x)$ for $a \leq x \leq b$), then the Riemann integral of $f(x)$ can be evaluated by the formula $\int_a^b f(x)\, dx = F(b) - F(a)$. A second form of this theorem states that if a function $f(x)$ is *continuous* on the interval $[a, b]$, then both (*i*) and (*ii*) of the preceding form are true, with $G(x) \equiv \int_a^x f(t)\, dt$ being a specific primitive, and for *any* primitive $F(x)$, $\int_a^b f(x)\, dx = F(b) - F(a)$. A third form of the theorem reads the same as the first form stated above, except that the function $F(x)$ is assumed merely to be continuous on $[a, b]$ and to possess a derivative $F'(x)$ equal to $f(x)$ at all but a finite number of points of $[a, b]$.

5. A Riemann-integrable function with a dense set of points of discontinuity.

Example 3 provides a monotonic function having the specified properties.

Example 15, Chapter 2, provides a nowhere monotonic function having the specified properties. In this latter case $\int_a^b f(x)\, dx = 0$ for all a and b.

6. A function f such that $g(x) \equiv \int_0^x f(t)\, dt$ is everywhere differentiable with a derivative different from $f(x)$ on a dense set.

If f is the function of Example 15, Chapter 2 (cf. the preceding

Example 5), $g(x) \equiv \int_0^x f(t)\,dt$ is identically zero, and therefore $g'(x) = 0$ for all x. Therefore $g'(x) = f(x)$ iff x is irrational.

7. Two distinct semicontinuous functions at a zero "distance."

In this case the **distance** d between two functions f and g integrable on $[a, b]$ is defined to be the integral of the absolute value of their difference:

$$d \equiv \int_a^b |f(x) - g(x)|\,dx.$$

If f is the semicontinuous function of the preceding example (cf. Example 16, Chapter 2) and if g is identically zero, then $f(x)$ and $g(x)$ are unequal for all rational values of x (and thus f and g are decidedly distinct functions), while the distance d defined above is equal to zero.

8. A Riemann-integrable function with an arbitrary F_σ set of measure zero as its set of points of discontinuity. (Cf. Example 22, Chapter 8.)

Somewhat as in Example 23, Chapter 2, let A be a given F_σ set of measure zero, $A = A_1 \cup A_2 \cup \cdots$, where $A_1, A_2, \cdots$ are closed subsets of an interval $[a, b]$ and $A_n \subset A_{n+1}$ for $n = 1, 2, \cdots$. Let A_0 denote the empty set $\emptyset$, and define the function f:

$$f(x) \equiv \begin{cases} 2^{-n} & \text{if } x \in A_n \setminus A_{n-1}, \\ 0 & \text{if } x \notin A. \end{cases}$$

If $c \in A$, f is discontinuous at c: if $c \in A_n \setminus A_{n-1}$, then since $A_n \setminus A_{n-1}$ is a set of measure zero it contains no interior points and c is a limit point of a set on which f has values differing from 2^{-n} by at least 2^{-n-1}. If $c \notin A$, f is continuous at c: if $\varepsilon > 0$ is given, choose N such that $2^{-N} < \varepsilon$; then choose a neighborhood of c that excludes $A_1, A_2, \cdots, A_n$ and inside which, therefore, $f(x) < 2^{-N} < \varepsilon$.

9. A Riemann-integrable function of a Riemann-integrable function that is not Riemann-integrable. (Cf. Example 34, Chapter 8.)

If $f(x) \equiv 1$ if $0 < x \leq 1$ and $f(0) \equiv 0$, and if g is the function f

of Example 15, Chapter 2, restricted to the closed interval [0, 1], then $f(g(x))$ is the characteristic function of the set Q of rational numbers, restricted to [0, 1], equal to 1 if x is rational and equal to 0 if x is irrational. (Cf. Example 1 of this chapter.)

10. A bounded monotonic limit of Riemann-integrable functions that is not Riemann-integrable. (Cf. Example 33, Chapter 8.)

The sequence $\{g_n\}$ defined in Example 24, Chapter 2, when restricted to the closed interval [0, 1], is an increasing sequence of Riemann-integrable functions; that is, for each $x \in [0, 1]$, $g_n(x) \leq g_{n+1}(x)$ for $n = 1, 2, \cdots$. If $g(x) \equiv \lim_{n \to +\infty} g_n(x)$ for $x \in [0, 1]$, then g is the characteristic function of the set Q of rational numbers, restricted to the closed interval [0, 1], and thus (cf. Example 1) g is not Riemann-integrable there.

11. A divergent improper integral that possesses a finite Cauchy principal value.

The improper integral $\int_{-\infty}^{+\infty} x \, dx$ is divergent, but its Cauchy principal value (cf. [34], p. 145, Ex. 30) is

$$\lim_{a \to +\infty} \int_{-a}^{a} x \, dx = \lim_{a \to +\infty} 0 = 0.$$

12. A convergent improper integral on [1, $+\infty$) whose integrand is positive, continuous, and does not approach zero at infinity.

For each integer $n > 1$ let $(gn) \equiv 1$, and on the closed intervals $[n - n^{-2}, n]$ and $[n, n + n^{-2}]$ define g to be linear and equal to 0 at the nonintegral endpoints. Finally, define $g(x)$ to be 0 for $x \geq 1$ where $g(x)$ is not already defined. Then the function

$$f(x) \equiv g(x) + \frac{1}{x^2}$$

is positive and continuous for $x \geq 1$, the statement $\lim_{x \to +\infty} f(x) = 0$ is false, and the improper integral

$$\int_{1}^{+\infty} f(x) \, dx$$

converges.

If the requirement of positivity is omitted, a simple example satisfying the remaining requirements (cf. [34], p. 146, Ex. 43) is $\int_1^{+\infty} \cos x^2 \, dx$.

13. A convergent improper integral on $[0, +\infty)$ whose integrand is unbounded in every interval of the form $[a, +\infty)$, where $a > 0$.

The improper integral $\int_0^{+\infty} x \cos x^4 \, dx$ satisfies these conditions (cf. [34], p. 146, Ex. 43).

An example where the integrand is everywhere positive and continuous can be constructed in a manner similar to that of the preceding Example 12 by letting $g(n) \equiv n$ and considering the closed intervals $[n - n^{-3}, n]$ and $[n, n + n^{-3}]$.

14. Functions f and g such that f is Riemann-Stieltjes integrable with respect to g on both $[a, b]$ and $[b, c]$, but not on $[a, c]$.

Let

$$f(x) \equiv \begin{cases} 0 & \text{if } 0 \leq x < 1, \\ 1 & \text{if } 1 \leq x \leq 2, \end{cases}$$

$$g(x) \equiv \begin{cases} 0 & \text{if } 0 \leq x \leq 1, \\ 1 & \text{if } 1 < x \leq 2, \end{cases}$$

and let $a = 0, b = 1$, and $c = 2$. Then

$$\int_0^1 f(x) \, dg(x) = 0, \qquad \int_1^2 f(x) \, dg(x) = 1,$$

but since f and g have a common point of discontinuity at $x = 1$,

$$\int_0^2 f(x) \, dg(x)$$

does not exist (cf. [34], p. 151, Ex. 10).

Chapter 5
Sequences

Introduction

The concepts of *sequence, Cauchy sequence, convergence,* and *divergence* are defined in the Introduction to Chapter 1. *Limits superior* and *inferior* at a (finite) point for functions are defined in the Introduction to Chapter 2. The corresponding formulations for sequences of real numbers will be assumed as known. The first six examples of the present chapter are concerned only with sequences of real numbers. For such sequences it should be emphasized that although the word *limit* is sometimes used in conjunction with the word *infinite*, the word *convergent* always implies a *finite* limit. It will be assumed (in Example 7) that the reader is familiar with the definition and elementary properties of *uniform convergence* of functions. Convergence and divergence for *sequences of sets* are defined with Example 8 for use with Examples 8 and 9. Throughout this book the single word *sequence* will be used to mean *infinite sequence* unless it is otherwise specifically modified.

1. Bounded divergent sequences.

The simplest example of a bounded divergent sequence is possibly

$$0. 1, 0, 1, \cdots ,$$

or $\{a_n\}$, where $a_n = 0$ if n is odd and $a_n = 1$ is n is even. Equivalently, $a_n = \frac{1}{2}(1 + (-1)^n)$.

A more extreme example is the sequence $\{r_n\}$ of rational numbers in $[0, 1]$ — that is, $\{r_n\}$ is a one-to-one correspondence with domain $\mathfrak{N}$ and range $\mathfrak{Q} \cap [0, 1]$.

2. For an arbitrary closed set, a sequence whose set of limit points is that set.

Any point that is the limit of a subsequence of a sequence $\{a_n\}$ is called a **limit point** or **subsequential limit** of the sequence. Any limit point of the range of a sequence is a limit point of the sequence, but the converse statement is not generally true. Counterexample: the alternating sequence 0, 1, 0, 1, $\cdots$ has two limit points, 0 and 1, but its range has none.

Since the set of all limit points of a sequence $\{a_n\}$ is the closure of the range of $\{a_n\}$, this set is always closed. The following example shows that *every* closed set A can be got in this way; in fact, that A is the set of limit points of a sequence $\{a_n\}$ of *distinct* points. It will follow that A is not only the set of limit points of the *sequence* $\{a_n\}$, but the set of limit points of its range as well.

If A is the empty set, let $a_n \equiv n$ for $n = 1, 2, \cdots$. Now let A be an arbitrary nonempty closed set (of real numbers), and let $\{r_n\}$ be an arrangement into a sequence of distinct terms of the set $\mathbb{Q}$ of all rational numbers ($\{r_n\}$ is a one-to-one correspondence with range $\mathbb{Q}$). The sequence $\{a_n\}$ whose set of limit points is A will be a subsequence of $\{r_n\}$ defined recursively as follows: We start by partitioning $\mathfrak{R}$ into the four disjoint intervals $(-\infty, -1)$, $[-1, 0)$, $[0, 1)$ and $[1, +\infty)$. If $A \cap (-\infty, -1) \neq \emptyset$, let a_1 be the first term of the sequence $\{r_n\}$ belonging to $(-\infty, -1)$; if $A \cap (-\infty, -1) = \emptyset$ and $A \cap [-1, 0) \neq \emptyset$, let a_1 be the first term $\in [-1, 0)$; if $A \cap (-\infty, 0) = \emptyset$, and $A \cap [0, 1) \neq \emptyset$, let a_1 be the first term $\in [0, 1)$; finally, if $A \cap (-\infty, 1) = \emptyset$, let a_1 be the first term $\in [1, +\infty)$. After a_1 is selected, a_2 distinct from a_1 is determined in like fashion by considering the intervals $[-1, 0)$, $[0, 1)$, and $[1, +\infty)$ in turn — unless $A \cap [-1, +\infty) = \emptyset$, in which case only a_1 is determined at this stage. In any case, at *least* one term a_1 and at *most* four terms a_1, a_2, a_3, a_4 of the sequence $\{a_n\}$ are thus defined. The second stage proceeds similarly, in terms of the partition of $\mathfrak{R}$ into the $2 \cdot 2^2 + 2 = 10$ intervals $(-\infty, -2)$, $[-2, -\frac{3}{2})$, $\cdots$, $[\frac{3}{2}, 2)$, $[2, +\infty)$. At each step, after a_1, a_2, $\cdots$, a_n are chosen, the term a_{n+1} is chosen from an interval I in case $A \cap I \neq \emptyset$, a_{n+1} being the first term of r_n *distinct from those already selected* and belonging to I. The kth set of $k \cdot 2^k + 2$ intervals consists of $(-\infty, -k)$, $[-k, -k + 2^{-k+1})$, $\cdots$, $[k - 2^{-k+1}, k)$, $[k, +\infty)$. It is not difficult to show that the sequence $\{a_n\}$, thus defined recursively, has the properties claimed. Notice that if $A = \mathfrak{R}$, then $\{a_n\}$ is a one-to-one correspondence with range $\mathbb{Q}$ — presumably distinct from $\{r_n\}$.

3. A divergent sequence $\{a_n\}$ for which $\lim\limits_{n\to+\infty} (a_{n+p} - a_n) = 0$ for every positive integer p.

Let a_n be the nth partial sum of the harmonic series:

$$a_n \equiv 1 + \tfrac{1}{2} + \cdots + \frac{1}{n}.$$

Then $\{a_n\}$ is divergent, but for $p > 0$,

$$a_{n+p} - a_n = \frac{1}{n+1} + \cdots + \frac{1}{n+p} \leq \frac{p}{n+1} \to 0.$$

It is important to note that the zero limit $\lim\limits_{n\to+\infty} (a_{n+p} - a_n) = 0$ is *not* uniform in p. In fact, for the stated properties to hold, this zero limit *cannot* be uniform in p since the statement that it *is* uniform in p is equivalent to the Cauchy criterion for convergence of a sequence (cf. [34], p. 447, Ex. 43).

One form of expressing the principal idea of the preceding paragraph is the following: If $\{a_n\}$ diverges, then there exists a strictly increasing sequence $\{p_n\}$ of positive integers such that $(a_{n+p_n} - a_n) \nrightarrow 0$. For the particular sequence of the partial sums of the harmonic series the sequence $\{p_n\}$ can be chosen to be $\{n\}$, since in this case

$$a_{n+p_n} - a_n = \frac{1}{n+1} + \cdots + \frac{1}{n+n} \geq \frac{n}{n+n} = \frac{1}{2}.$$

The following example is related to another aspect of this question (with $\phi(n) = n + p_n$).

4. For an arbitrary strictly increasing sequence $\{\phi_n\} = \{\phi(n)\}$ of positive integers, a divergent sequence $\{a_n\}$ such that $\lim\limits_{n\to+\infty} (a_{\phi(n)} - a_n) = 0$.

By induction, $\phi(n) \geq n$ for all $n = 1, 2, \cdots$, and more generally, $\phi(n + k) \geq n + \phi(k)$ for all n and $k = 1, 2, \cdots$. Therefore $\lim\limits_{n\to+\infty} \phi(n) = +\infty$. There are two cases to consider.

If $\phi(n) - n$ is bounded, say $\phi(n) - n \leq K$ for all $n = 1, 2, \cdots$, then the sequence $\{a_n\}$ can be chosen to be the sequence of partial sums of the harmonic series, since

$$a_{\phi(n)} - a_n = \frac{1}{n+1} + \cdots + \frac{1}{\phi(n)} \leq \frac{K}{n+1} \to 0.$$

If $\phi(n) - n$ is unbounded, let k be the smallest positive integer such that $\phi(k) > k$, and define a_n to be equal to 1 if $n = k$, $\phi(k)$, $\phi(\phi(k))$, $\cdots$, and equal to 0 otherwise. Since $\{\phi(n)\}$ is strictly increasing there exists a subsequence of $\{a_n\}$ identically equal to 1, and since $\phi(n) - n$ is unbounded there exists a subsequence of $\{a_n\}$ identically equal to 0. Therefore $\{a_n\}$ diverges. On the other hand, $a_{\phi(n)} = a_n$ for every $n = 1, 2, \cdots$, and therefore $a_{\phi(n)} - a_n \to 0$.

This example can be generalized in various ways. For example, it is sufficient to assume merely that $\phi(n) \to +\infty$ as $n \to +\infty$, and it is possible at the same time to require that $\{a_n\}$ be unbounded. Space does not permit inclusion of the details.

5. Sequences $\{a_n\}$ and $\{b_n\}$ such that $\underline{\lim}\ a_n + \underline{\lim}\ b_n < \underline{\lim}\ (a_n + b_n) < \underline{\lim}\ a_n + \overline{\lim}\ b_n < \overline{\lim}\ (a_n + b_n) < \overline{\lim}\ a_n + \overline{\lim}\ b_n$.

Let $\{a_n\}$ and $\{b_n\}$ be the sequences repeating in cycles of 4:

$$\{a_n\}: 0, 1, 2, 1, 0, 1, 2, 1, 0, 1, 2, 1, 0, 1, 2, 1, \cdots,$$

$$\{b_n\}: 2, 1, 1, 0, 2, 1, 1, 0, 2, 1, 1, 0, 2, 1, 1, 0, \cdots.$$

Then the inequality statement specified above becomes

$$0 < 1 < 2 < 3 < 4.$$

6. Sequences $\{a_{1n}\}$, $\{a_{2n}\}$, $\cdots$ such that

$$\overline{\lim}_{n \to +\infty} (a_{1n} + a_{2n} + \cdots) > \overline{\lim}_{n \to +\infty} a_{1n} + \overline{\lim}_{n \to +\infty} a_{2n} + \cdots.$$

Such an example is given by $a_{mn} = 1$ if $m = n$ and $a_{mn} = 0$ if $m \neq n$, $m, n = 1, 2, \cdots$, where all infinite series involved converge. The inequality stated above becomes $1 > 0$.

It should be noted that the inequality exhibited in this example is impossible if there are only finitely many sequences. For example,

$$\varlimsup_{n \to +\infty} (a_n + b_n) \leqq \varlimsup_{n \to +\infty} a_n + \varlimsup_{n \to +\infty} b_n.$$

(Cf. [34], p. 59, Ex. 19.)

7. Two uniformly convergent sequences of functions the sequence of whose products does not converge uniformly.

On any common domain D let f be any unbounded function and let the sequences $\{f_n\}$ and $\{g_n\}$ be defined:

$$f_n(x) \equiv f(x), \qquad g_n(x) \equiv 1/n.$$

Then $f_n \to f$ and $g_n \to 0$ uniformly on D , but $f_n g_n \to 0$ nonuniformly on D. A specific example is $D = \mathfrak{R}, f(x) = x$.

It should be noted that if both sequences are bounded and converge uniformly on D, then the product sequence also converges uniformly on D.

8. A divergent sequence of sets.

The **limit superior** and **limit inferior** of a sequence $\{A_n\}$ of sets are defined and denoted:

$$\varlimsup_{n \to +\infty} A_n \equiv \bigcap_{n=1}^{+\infty} [\bigcup_{m=n}^{+\infty} A_m], \qquad \varliminf_{n \to +\infty} A_n \equiv \bigcup_{n=1}^{+\infty} [\bigcap_{m=n}^{+\infty} A_m],$$

respectively. A sequence $\{A_n\}$ is **convergent** iff $\varlimsup_{n \to +\infty} A_n = \varliminf_{n \to +\infty} A_n$ and, in this case, converges to this common value. A sequence of sets is **divergent** iff it fails to be convergent. Since $\varlimsup_{n \to +\infty} A_n = \{x \mid x \in$ infinitely many $A_n\}$ and $\varliminf_{n \to +\infty} A_n = \{x \mid x \in$ all but a finite number of $A_n\}$, the oscillating sequence $A, B, A, B, A, B, \cdots$ has limit superior equal to the union $A \cup B$ and limit inferior equal to the intersection $A \cap B$. Such a sequence therefore converges iff $A = B$.

The close analogy between this example and the alternating sequence $\{a, b, a, b, \cdots\}$ of numbers (cf. Example 1, above) should not escape notice.

9. A sequence $\{A_n\}$ of sets that converges to the empty set but whose cardinal numbers $\to +\infty$.

Let A_n be defined to be the set of n positive integers greater than or equal to n and less than $2n$:

$$A_n \equiv \{m \mid m \in \mathfrak{N}, n \leq m < 2n\}, \qquad n = 1, 2, \cdots.$$

Then, since no positive integer belongs to infinitely many of the sets of $\{A_n\}$, the limit superior and limit inferior are both empty.

The preceding example can be visualized in terms of placing pairs of billiard balls, which bear numbers 0, 1, 2, $\cdots$, into a bag while repeatedly withdrawing one. For example, at one minute before noon balls numbered 0 and 1 are placed in the bag and ball number 0 is removed. At $\frac{1}{2}$ minute before noon balls numbered 2 and 3 are added, and ball number 1 is removed. At $\frac{1}{3}$ minute before noon balls 4 and 5 are added and ball number 2 is taken out. This process is continued, and the question is asked: "How many balls are in the bag at noon?" Answer: "None."

Since the natural numbers can be put into a one-to-one correspondence with their reciprocals, and since as subsets of $\mathfrak{R}$ all finite sets are compact (closed and bounded), all of the sets A_n of this example are compact, and can even be assumed to be *uniformly* bounded (contained in the same bounded interval). If the sequence $\{A_n\}$ is assumed to be *decreasing* ($A_{n+1} \subset A_n$ for $n = 1, 2, \cdots$), then the limit $\lim_{n \to +\infty} A_n$ is the same as the intersection $\bigcap\limits_{n=1}^{+\infty} A_n$, and may be empty even though the cardinal number of every A_n is infinite and even though every A_n is bounded $\left(\text{example:} \left\{ \dfrac{1}{n}, \dfrac{1}{n+1}, \cdots \right\} \right)$ or closed (example: $\{n, n+1, \cdots\}$) but *not both*. (Cf. [34], p. 201.)

Chapter 6
Infinite Series

Introduction

Unless explicit statement to the contrary is made, all series considered in this chapter will be assumed to be real, that is, consisting of terms that are real numbers. If $\{s_n\}$ is the sequence of partial sums of an infinite series $\sum a_n = a_1 + a_2 + \cdots + a_n + \cdots$, that is, $s_n = a_1 + \cdots + a_n$ for $n = 1, 2, \cdots$, the series $\sum a_n$ is said to **converge** iff $\lim_{n \to +\infty} s_n$ exists and is finite. This limit s is called the **sum** of the series $\sum a_n$, with the alternative notations:

$$\sum a_n = \sum_{n=1}^{+\infty} a_n = s.$$

The series $\sum a_n$ is said to **diverge** iff it fails to converge, that is, iff $\lim_{n \to +\infty} a_n$ is infinite or fails to exist. The statement $\sum a_n = +\infty$ means that $\lim_{n \to +\infty} s_n = +\infty$. A sequence $\{a_n\}$ or a series $\sum a_n$ is **nonnegative** or **positive** iff $a_n \geq 0$ for every n or $a_n > 0$ for every n, respectively. For a nonnegative series $\sum a_n$, the statement $\sum a_n < +\infty$ means that the series converges, and the statement $\sum a_n = +\infty$ means that the series diverges.

For certain purposes series may start with a term a_0, in which case $\sum a_n$ should be interpreted to mean $\sum_{n=0}^{+\infty} a_n$, or the sum of this series. For a power series $\sum_{n=0}^{+\infty} a_n x^n$, the term $a_0 x^0$ should be understood to mean a_0 even when $x = 0$; that is, for present purposes $0^0 \equiv 1$. For a Maclaurin series

$$\sum_{n=0}^{+\infty} \frac{f^{(n)}(0)}{n!} x^n,$$

the term for $n = 0$ is $f(0)$; in other words, $f^{(0)}(x) \equiv f(x)$.

1. A divergent series whose general term approaches zero.
The harmonic series $\sum 1/n$.

2. A convergent series $\sum a_n$ and a divergent series $\sum b_n$ such that $a_n \geqq b_n$, $n = 1, 2, \cdots$.
Let $a_n \equiv 0$ and $b_n \equiv -1/n$, $n = 1, 2, \cdots$.

3. A convergent series $\sum a_n$ and a divergent series $\sum b_n$ such that $|a_n| \geqq |b_n|$, $n = 1, 2, \cdots$.
Let $\sum a_n$ be the conditionally convergent alternating harmonic series $\sum (-1)^{n+1}/n$, and let $\sum b_n$ be the divergent harmonic series $\sum 1/n$.

4. For an arbitrary given positive series, either a dominated divergent series or a dominating convergent series.
A nonnegative series $\sum a_n$ is said to **dominate** a series $\sum b_n$ iff $a_n \geqq |b_n|$ for $n = 1, 2, \cdots$. If the given positive series is $\sum b_n$, let $a_n \equiv b_n$ for $n = 1, 2, \cdots$. Then if $\sum b_n$ diverges it dominates the divergent series $\sum a_n$, and if $\sum b_n$ converges it is dominated by the convergent series $\sum a_n$. The domination inequalities can all be made strict by means of factors $\frac{1}{2}$ and 2.

This simple result can be framed as follows: *There exists no positive series that can serve simultaneously as a comparison test series for convergence and as a comparison test series for divergence.* (Cf. Example 19, below.)

5. A convergent series with a divergent rearrangement.
With any conditionally convergent series $\sum a_n$, such as the alternating harmonic series $\sum (-1)^{n+1}/n$, the terms can be rearranged in such a way that the new series is convergent to any given sum, or is divergent. Divergent rearrangements can be found so that the sequence $\{s_n\}$ of partial sums has the limit $+\infty$, the limit $-\infty$, or no limit at all. In fact, the sequence $\{s_n\}$ can be determined in such a way that its set of limits points is an arbitrary given closed interval, bounded or not (cf. Example 2, Chapter 5). The underlying reason that this is possible is that the series of positive terms of $\sum a_n$ and the series of negative terms of $\sum a_n$ are both divergent.

To be specific we shall indicate a rearrangement of the series $\sum(-1)^{n+1}/n$ such that the sequence $\{s_n\}$ of partial sums has the closed interval $[a, b]$ as its set of limit points. We start with the single term 1, and then attach negative terms:

$$1 - \frac{1}{2} - \frac{1}{4} - \frac{1}{6} - \frac{1}{8} - \cdots - \frac{1}{2j}$$

until the sum first is less than a. Then we add on unused positive terms:

$$1 - \frac{1}{2} - \frac{1}{4} - \cdots - \frac{1}{2j} + \frac{1}{3} + \frac{1}{5} + \cdots + \frac{1}{2k + 1}$$

until the sum first is greater than b. Continuing with this idea we adjoin negative terms until the sum first is less than a; then positive terms until the sum first exceeds b; then negative terms, etc., *ad infinitum*. Since the absolute value $1/n$ of the general term $(-1)^{n+1}/n$ approaches zero, it follows that every number of the closed interval $[a, b]$ is approached arbitrarily closely by partial sums s_n of the rearranged series, for arbitrarily large n. Furthermore, for no number outside the interval $[a, b]$ is this true.

In the procedure just described, if the partial sums are permitted to go just above 1, then just below -2, then just above 3, then just below -4, etc., the sequence of partial sums of the rearranged series has the entire real number system as its set of limit points.

W. Sierpinski (cf. [43]) has shown that if $\sum a_n$ is a conditionally convergent series with sum s, and if $s' < s$, then for some rearrangement involving the positive terms only (leaving the negative terms in their original positions) the rearranged series has sum s'. A similar remark applies to numbers $s'' > s$ and rearrangements involving only negative terms. This is clearly an extension of the celebrated "Riemann derangement theorem" (cf. [36], p. 232, Theorem III), illustrated in all its essentials by the discussion in Example 5.

In a different direction, there is an extension that reads: If $\sum a_n$ is a conditionally convergent series of *complex* numbers then the sums obtainable by all possible rearrangements that are either convergent or divergent to ∞ constitute a set that is either a single line in the complex plane (including the point at infinity) or the complex plane *in toto* (including the point at infinity). Furthermore, if $\sum v_n$ is a conditionally convergent series of *vectors* in a finite-dimensional space, then the sums obtainable by all possible rearrangements constitute a set that is some linear variety in the space (cf. [47]).

6. For an arbitrary conditionally convergent series $\sum a_n$ and an arbitrary real number x, a sequence $\{\varepsilon_n\}$, where $|\varepsilon_n| = 1$ for $n = 1, 2, \cdots$, such that $\sum \varepsilon_n a_n = x$.

The procedure here is similar to that employed in Example 5. Since $\sum |a_n| = +\infty$, we may attach factors ε_n of absolute value 1 in such a fashion that $\varepsilon_1 a_1 + \cdots + \varepsilon_n a_n = |a_1| + \cdots + |a_n| > x$. Let n_1 be the least value of n that ensures this inequality. We then provide factors ε_n, of absolute value 1, for the next terms in order to obtain (for the least possible n_2):

$$\varepsilon_1 a_1 + \cdots + \varepsilon_{n_2} a_{n_2} = |a_1| + \cdots + |a_{n_1}|$$
$$- |a_{n_1+1}| - \cdots - |a_{n_2}| < x.$$

If this process is repeated, with partial sums alternately greater than x and less than x, a series $\sum \varepsilon_n a_n$ is obtained which, since $a_n \to 0$ as $n \to +\infty$, must converge to x.

7. Divergent series satisfying any two of the three conditions of the standard alternating series theorem.

The alternating series theorem referred to states that the series $\sum \varepsilon_n c_n$, where $|\varepsilon_n| = 1$ and $c_n > 0$, $n = 1, 2, \cdots$, converges provided

    (*i*) $\varepsilon_n = (-1)^{n+1}$, $n = 1, 2, \cdots$,

    (*ii*) $c_{n+1} \leqq c_n$, $n = 1, 2, \cdots$,

    (*iii*) $\lim_{n \to +\infty} c_n = 0$.

No two of these three conditions by themselves imply convergence; that is, no one can be omitted. The following three examples demonstrate this fact:

(*i*): Let $\varepsilon_n \equiv 1$, $c_n \equiv 1/n$, $n = 1, 2, \cdots$. Alternatively, for an example that is, after a fashion, an "alternating series" let $\{\varepsilon_n\}$ be the sequence repeating in triplets: $1, 1, -1, 1, 1, -1, \cdots$.

(*ii*): Let $c_n \equiv 1/n$ if n is odd, and let $c_n \equiv 1/n^2$ if n is even.

(*iii*): Let $c_n \equiv (n + 1)/n$ (or, more simply, let $c_n \equiv 1$), $n = 1, 2, \cdots$.

8. A divergent series whose general term approaches zero and which, with a suitable introduction of parentheses, becomes convergent to an arbitrary sum.

Introduction of parentheses in an infinite series means grouping of consecutive finite sequences of terms (each such finite sequence consisting of at least one term) to produce a new series, whose sequence of partial sums is therefore a subsequence of the sequence of partial sums of the original series. For example, one way of introducing parentheses in the alternating harmonic series gives the series

$$\left(1 - \frac{1}{2}\right) + \left(\frac{1}{3} - \frac{1}{4}\right) + \cdots$$
$$= \frac{1}{1\cdot 2} + \frac{1}{3\cdot 4} + \cdots + \frac{1}{(2n - 1)\cdot 2n} + \cdots.$$

Any series derived from a convergent series by means of introduction of parentheses is convergent, and has a sum equal to that of the given series.

The final rearranged series described under Example 5 has the stated property since, for an arbitrary real number, a suitable introduction of parentheses gives a convergent series whose sum is the given number.

9. For a given positive sequence $\{b_n\}$ with limit inferior zero, a positive divergent series $\sum a_n$ whose general term approaches zero and such that $\underline{\lim}_{n\to+\infty} a_n/b_n = 0$.

Choose a subsequence $b_{n_1}, b_{n_2}, \cdots, b_{n_k}, \cdots$ of nonconsecutive terms of $\{b_n\}$ such that $\lim_{k\to+\infty} b_{n_k} = 0$, and let $a_{n_k} \equiv b_{n_k}^2$ for $k = 1, 2, \cdots$. For every *other* value of $n : n = m_1, m_2, m_3, \cdots, m_j, \cdots$, let $a_{m_j} \equiv 1/j$. Then $a_n \to 0$ as $n \to +\infty$, $\sum a_n$ diverges, and $a_{n_k}/b_{n_k} = b_{n_k} \to 0$ as $k \to +\infty$.

This example shows (in particular) that no matter how rapidly a positive sequence $\{b_n\}$ may converge to zero, there is a positive sequence $\{a_n\}$ that converges to zero slowly enough to ensure divergence of the series $\sum a_n$, and yet has a subsequence converging to zero more rapidly than the corresponding subsequence of $\{b_n\}$.

10. For a given positive sequence $\{b_n\}$ with limit inferior zero, a positive convergent series $\sum a_n$ such that $\overline{\lim}_{n\to+\infty} a_n/b_n = +\infty$.

Choose a subsequence $b_{n_1}, b_{n_2}, \cdots, b_{n_k}, \cdots$ of $\{b_n\}$ such that for each positive integer k, $b_{n_k} < k^{-3}$, and let $a_{n_k} \equiv k^{-2}$ for $k = 1, 2, \cdots$. For every other value of n let $a_n \equiv n^{-2}$. Then $\sum a_n < +\infty$, while $a_{n_k}/b_{n_k} = k \to +\infty$.

This example shows (in particular) that no matter how slowly a positive sequence $\{b_n\}$ may converge to zero, there is a sequence $\{a_n\}$ of positive numbers that converges to zero rapidly enough to ensure convergence of the series $\sum a_n$, and yet has a subsequence converging to zero more slowly than the corresponding subsequence of $\{b_n\}$.

11. For a positive sequence $\{c_n\}$ with limit inferior zero, a positive convergent series $\sum a_n$ and a positive divergent series $\sum b_n$ such that $a_n/b_n = c_n$, $n = 1, 2, \cdots$.

Choose a subsequence $c_{n_1}, c_{n_2}, \cdots, c_{n_k}, \cdots$ of $\{c_n\}$ such that for each positive integer k, $c_{n_k} < k^{-2}$, and let $a_{n_k} \equiv c_{n_k}$, $b_{n_k} \equiv 1$ for $k = 1, 2, \cdots$. For every other value of n let $a_n \equiv n^{-2}$, $b_n \equiv (n^2 c_n)^{-1}$. Then $\sum a_n$ converges, $\sum d_n$ diverges since $b_n \nrightarrow 0$ as $n \to +\infty$, and $a_n/b_n = c_n$ for $n = 1, 2, \cdots$.

This example shows (in particular) that no matter how slowly a positive sequence $\{c_n\}$ may converge to zero, there exist positive series of which one is convergent and the other is divergent, the quotient of whose nth terms is c_n.

12. A function positive and continuous for $x \geq 1$ and such that $\int_1^{+\infty} f(x)dx$ converges and $\sum_{n=1}^{+\infty} f(n)$ diverges.
Example 12, Chapter 4.

13. A function positive and continuous for $x \geq 1$ and such that $\int_1^{+\infty} f(x)dx$ diverges and $\sum_{n=1}^{+\infty} f(n)$ converges.

For each $n > 1$ let $g(n) \equiv 0$, and on the closed intervals $[n - n^{-1}, n]$ and $[n, n + n^{-1}]$ define g to be linear and equal to 1 at the nonintegral endpoints. Finally, define $g(x)$ to be 1 for $x \geq 1$ where $g(x)$ is not already defined. Then the function

$$f(x) \equiv g(x) + \frac{1}{x^2}$$

is positive and continuous for $x \geq 1$, $\int_1^{+\infty} f(x)dx = +\infty$, and $\sum_{n=1}^{+\infty} f(n) = \sum_{n=1}^{+\infty} n^{-2} < +\infty$.

14. Series for which the ratio test fails.

For a positive series $\sum a_n$ the ratio test states (cf. [34], p. 390) that if the limit

$$\lim_{n \to +\infty} \frac{a_{n+1}}{a_n} = \rho$$

exists in the finite or infinite sense ($0 \le \rho \le +\infty$), then
  (i) if $0 \le \rho < 1$, $\sum a_n$ converges;
  (ii) if $1 < \rho \le +\infty$, $\sum a_n$ diverges.

The statement that if $\rho = 1$ the test fails is more than an empty statement. It means that there exist both convergent and divergent positive series for each of which $\rho = 1$. Examples are

$$\sum_{n=1}^{+\infty} \frac{1}{n^2} \quad \text{and} \quad \sum_{n=1}^{+\infty} \frac{1}{n},$$

respectively.

The ratio test may also fail by virtue of the nonexistence of the limit ρ. Examples of convergent and divergent positive series are respectively

$$\sum_{n=1}^{+\infty} 2^{(-1)^n - n} = \frac{1}{2^2} + \frac{1}{2^1} + \frac{1}{2^4} + \frac{1}{2^3} + \frac{1}{2^6} + \frac{1}{2^5} + \cdots,$$

where

$$\overline{\lim_{n \to +\infty}} \frac{a_{n+1}}{a_n} = 2 \quad \text{and} \quad \underline{\lim_{n \to +\infty}} \frac{a_{n+1}}{a_n} = \frac{1}{8},$$

and

$$\sum_{n=1}^{+\infty} 2^{n - (-1)^n} = 2^2 + 2^1 + 2^4 + 2^3 + 2^6 + 2^5 + \cdots,$$

where

$$\overline{\lim_{n \to +\infty}} \frac{a_{n+1}}{a_n} = 8 \quad \text{and} \quad \underline{\lim_{n \to +\infty}} \frac{a_{n+1}}{a_n} = \frac{1}{2}.$$

A refined form of the ratio test states that if

(iii) $\overline{\lim_{n \to +\infty}} \dfrac{a_{n+1}}{a_n} < 1$, $\sum a_n$ converges;

(iv) $\underline{\lim_{n \to +\infty}} \dfrac{a_{n+1}}{a_n} > 1$, $\sum a_n$ diverges.

This form of the ratio test may fail as a result of the inequalities

$$\varliminf_{n \to +\infty} \frac{a_{n+1}}{a_n} \leqq 1 \leqq \varlimsup_{n \to +\infty} \frac{a_{n+1}}{a_n} .$$

Examples for which both equalities and strict inequalities occur are given above.

15. Series for which the root test fails.

For a nonnegative series $\sum a_n$ the simplest form of the root test states (cf. [34], p. 392) that if

$$\lim_{n \to +\infty} \sqrt[n]{a_n} = \sigma$$

exists in the finite or infinite sense ($0 \leqq \sigma \leqq +\infty$), then

(*i*) if $0 \leqq \sigma < 1$, $\sum a_n$ converges;

(*ii*) if $1 < \sigma \leqq +\infty$, $\sum a_n$ diverges.

The ratio test and the root test are related by the fact that if $\lim_{n \to +\infty} (a_{n+1}/a_n)$ exists in the finite or infinite sense, then $\lim_{n \to +\infty} \sqrt[n]{a_n}$ exists and is equal to it (cf. [34], p. 394, Ex. 31). Consequently, if the ratio test in the form (*i*) or (*ii*), Example 14, is successful, the root test is also. Furthermore, the first two examples illustrating failure of the ratio test (Example 14) also serve to exemplify failure of the root test for the same reason. The last two examples of the failure of the ratio test illustrate the possibility of success for the root test (cf. Example 16).

The root test, as stated above, fails for the convergent series

$$\sum_{n=1}^{+\infty} \left(\frac{5 + (-1)^n}{2} \right)^{-n} = \frac{1}{2} + \frac{1}{3^2} + \frac{1}{2^3} + \frac{1}{3^4} + \frac{1}{2^5} + \frac{1}{3^6} + \cdots ,$$

since $\varlimsup_{n \to +\infty} \sqrt[n]{a_n} = \frac{1}{2}$ and $\varliminf_{n \to +\infty} \sqrt[n]{a_n} = \frac{1}{3}$. It also fails for the divergent series

$$\sum_{n=1}^{+\infty} \left(\frac{5 + (-1)^n}{2} \right)^n ,$$

since $\varlimsup_{n \to +\infty} \sqrt[n]{a_n} = 3$ and $\varliminf_{n \to +\infty} \sqrt[n]{a_n} = 2$.

A refined form of the root test states that if

(*iii*) $\varlimsup_{n \to +\infty} \sqrt[n]{a_n} < 1$, $\sum a_n$ converges;

(*iv*) $\varlimsup_{n \to +\infty} \sqrt[n]{a_n} > 1$, $\sum a_n$ diverges.

This form of the root test is at least as strong as (actually stronger than — cf. Example 16) the refined form of the ratio test (Example 14), since (cf. [34], p. 394, Ex. 31)

$$\varliminf_{n \to +\infty} \frac{a_{n+1}}{a_n} \leq \varliminf_{n \to +\infty} \sqrt[n]{a_n} \leq \varlimsup_{n \to +\infty} \sqrt[n]{a_n} \leq \varlimsup_{n \to +\infty} \frac{a_{n+1}}{a_n}.$$

The refined form of the root test establishes convergence and divergence for the preceding two examples, since $\varlimsup_{n \to +\infty} \sqrt[n]{a_n}$ is equal to $\frac{1}{2}$ and 3, respectively, for these two series.

The refined form of the root test can fail only by virtue of the equation $\varlimsup_{n \to +\infty} \sqrt[n]{a_n} = 1$. Examples of this type of failure have already been given.

16. Series for which the root test succeeds and the ratio test fails.

The convergent series $\sum_{n=1}^{+\infty} 2^{(-1)^n - n}$ for which the ratio test fails (Example 14) is one for which the root test succeeds. Indeed,

$$\sqrt[n]{a_n} = 2^{\frac{(-1)^n - n}{n}} \to 2^{-1} = \tfrac{1}{2} < 1.$$

Similarly, for the divergent series $\sum_{n=0}^{+\infty} 2^{n - (-1)^n}$, of Example 14,

$$\sqrt[n]{a_n} = 2^{\frac{n - (-1)^n}{n}} \to 2^1 = 2 > 1.$$

17. Two convergent series whose Cauchy product series diverges.

The **Cauchy product series** of the two series $\sum_{n=0}^{+\infty} a_n$ and $\sum_{n=0}^{+\infty} b_n$ is defined to be the series $\sum_{n=0}^{+\infty} c_n$, where

$$c_n = \sum_{k=0}^{n} a_k b_{n-k} = a_0 b_n + a_1 b_{n-1} + \cdots + a_n b_0.$$

The theorem of Mertens (cf. [36], p. 239, Ex. 20) states that if $\sum a_n$ converges to A, if $\sum b_n$ converges to B, and if *one* at least of these convergences is *absolute*, then the product series $\sum c_n$ converges to AB.

Let $\sum a_n$ and $\sum b_n$ be the identical series

$$a_n = b_n = (-1)^n (n + 1)^{-1/2}, \qquad n = 0, 1, 2, \cdots.$$

Then $\sum a_n$ and $\sum b_n$ converge by the alternating series test (Example 7), while $\sum c_n$ diverges since

$$|c_n| = \frac{1}{\sqrt{1}\sqrt{n+1}} + \frac{1}{\sqrt{2}\sqrt{n}} + \frac{1}{\sqrt{3}\sqrt{n-1}} + \cdots$$
$$+ \frac{1}{\sqrt{n+1}\sqrt{1}}$$
$$\geqq \frac{2}{n+2} + \frac{2}{n+2} + \frac{2}{n+2} + \cdots + \frac{2}{n+2}$$
$$= \frac{2(n+1)}{n+2} \geqq 1, \qquad n = 0,1,2,\cdots,$$

the first inequality holding since $\sqrt{(1+x)(n+1-x)}$ on the closed interval $[0, n]$ is maximized when $x = \frac{1}{2}n$.

18. Two divergent series whose Cauchy product series converges absolutely.

The Cauchy product series of the two series

$$2 + 2 + 2^2 + 2^3 + \cdots + 2^n + \cdots, \qquad n = 1, 2, \cdots,$$
$$-1 + 1 + 1 + 1 + \cdots + 1^n + \cdots, \qquad n = 1, 2, \cdots,$$

is

$$-2 + 0 + 0 + 0 + \cdots + 0^n + \cdots, \qquad n = 1, 2, \cdots.$$

More generally, if $a_n = a^n$ for $n \geqq 1$ and if $b_n = b^n$ for $n \geqq 1$, and if $a \neq b$, the term c_n of the product series of $\sum a_n$ and $\sum b_n$, for $n \geqq 1$, is

$$c_n = a_0 b^n + b_0 a^n + a^{n-1} b + a^{n-2} b^2 + a^{n-3} b^3 + \cdots ab^{n-1}$$
$$= a_0 b^n + b_0 a^n - a^n - b^n + (a^{n+1} - b^{n+1})/(a - b)$$
$$= \{a^n[a + (b_0 - 1)(a - b)] - b^n[b + (1 - a_0)(a - b)]\}/(a - b),$$

and therefore $c_n = 0$ in case $a = (1 - b_0)(a - b)$ and $b = (a_0 - 1)(a - b)$. If a and b are chosen so that $a - b = 1$, then a_0 and b_0 are given by $a_0 = b + 1 = a$, $b_0 = 1 - a = -b$.

19. For a given sequence $\{\sum_{m=1}^{+\infty} a_{mn}\}$, $n = 1, 2, \cdots$, of positive convergent series, a positive convergent series $\sum_{m=1}^{+\infty} a_m$ that does not compare favorably with any series of $\{\sum_{m=1}^{+\infty} a_{mn}\}$.

The statement that the series $\sum_{m=1}^{+\infty} a_m$ **compares favorably** with the series $\sum_{m=1}^{+\infty} a_{mn}$, for a fixed positive integer n, means:

$$\exists\ M \in \mathfrak{N} \ni m > M \Rightarrow a_m \leqq a_{mn},$$

and therefore the statement that such comparison does *not* obtain means:

$$\forall\ M \in \mathfrak{N} \,\exists\, m > M \ni a_m > a_{mn}.$$

For all positive integers M and n, define the *positive numbers* S_n, S_{Mn}, and R_{Mn}:

$$S_n \equiv \sum_{m=1}^{+\infty} a_{mn}, \qquad S_{Mn} \equiv \sum_{m=1}^{M} a_{mn}, \qquad R_{Mn} \equiv \sum_{m=M+1}^{+\infty} a_{mn},$$

and for each $n \in \mathfrak{N}$ choose $M(n)$ so that $1 \leqq M(1) < M(2) < \cdots$ and

$$R_{M(1),1} < 2^{-1},$$

$$\max\,(R_{M(2),1}\,,\,R_{M(2),2}) < 2^{-2},$$

$$\cdots$$

$$\max\,(R_{M(n),1}\,,\,\cdots\,,\,R_{M(n),n}) < 2^{-n}.$$

For any positive integer m, define a_m as follows:

$$a_m \equiv \begin{cases} 2a_{m1} & \text{if } 1 \leqq m \leqq M(2), \\ (k+1)\max\,(a_{m1},\,a_{m2},\,\cdots\,,\,a_{mk}) \\ \qquad \text{if } M(k) < m \leqq M(k+1) \text{ for } k > 1. \end{cases}$$

In order to prove that $\sum a_m$ converges, we first establish an inequality for the (finite) sum of the terms of this series for $M(k) < m \leqq M(k+1)$, where $k > 1$:

$$\sum_{m=M(k)+1}^{M(k+1)} a_m \leqq \sum_{m=M(k)+1}^{M(k+1)} \left[(k+1) \sum_{n=1}^{k} a_{mn} \right]$$

$$= (k+1) \sum_{n=1}^{k} \left[\sum_{m=M(k)+1}^{M(k+1)} a_{mn} \right] \leqq (k+1) \sum_{n=1}^{k} R_{M(k),\,n}$$

$$\leqq (k+1) \sum_{n=1}^{k} 2^{-k} < (k+1)^2 2^{-k}.$$

We therefore have

$$\sum_{m=1}^{+\infty} a_m = \sum_{m=1}^{M(2)} a_m + \sum_{m=M(2)+1}^{M(3)} a_m \cdots$$

$$\leqq 2 \sum_{m=1}^{M(2)} a_{m1} + \sum_{k=2}^{+\infty} \sum_{m=M(k)+1}^{M(k+1)} a_m$$

$$< 2S_{M(2),1} + \sum_{k=2}^{+\infty} (k+1)^2 \, 2^{-k} < +\infty.$$

On the other hand, for any fixed n, whenever $k \geqq n$ and $m > M(k)$, $a_m/a_{mn} \geqq k+1$, whence $\lim_{m \to +\infty} a_m/a_{mn} = +\infty$. Therefore the series $\sum_{m=1}^{+\infty} a_m$ does not compare favorably with the series $\sum_{m=1}^{+\infty} a_{mn}$. In fact, the series $\sum_{m=1}^{+\infty} a_m$ does not compare favorably with the series $\sum_{m=1}^{+\infty} a_{mn}$ even when such favorable comparison is defined:

$$\exists \, M \text{ and } K \in \mathfrak{N} \ni m > M \Rightarrow a_m \leqq K a_{mn},$$

or equivalently, when failure to have favorable comparison means:

$$\forall \, M \text{ and } K \in \mathfrak{N} \, \exists \, m > M \ni a_m > K a_{mn}.$$

A sequence of positive convergent series is called a **universal comparison sequence** if and only if it has the property that a given positive series converges *if and only if* it compares favorably with at least one series of the universal comparison sequence. That is, a sequence of positive convergent series is a universal comparison sequence if and only if the convergence or divergence of every positive series can be established by comparison with some member series of the sequence. Example 19 shows that *no such universal comparison sequence exists*.

20. A Toeplitz matrix T and a divergent sequence that is transformed by T into a convergent sequence.

An **infinite matrix** is a real-valued or complex-valued function on $\mathfrak{N} \times \mathfrak{N}$, denoted $T = (t_{ij})$, where i and $j \in \mathfrak{N}$. In case the infinite series $\sum_{j=1}^{+\infty} t_{ij} a_j$, where $\{a_j\}$ is a given sequence of numbers, converges for every $i \in \mathfrak{N}$, the sequence $\{b_i\}$, defined by

$$b_i \equiv \sum_{j=1}^{+\infty} t_{ij} a_j,$$

is called the **transform** of $\{a_j\}$ by T. An infinite matrix $T = (t_{ij})$

is called a **Toeplitz matrix**[†] iff for every convergent sequence $\{a_j\}$ the sequence $\{b_i\}$ is well defined, and the limit $\lim_{i \to +\infty} b_i$ exists and is equal to the limit $\lim_{j \to +\infty} a_j$. An important basic fact in the theory of Toeplitz matrices is that the following three properties form a necessary and sufficient set of conditions for an infinite matrix $T = (t_{ij})$ to be a Toeplitz matrix (for a proof see [49]):

(1) $$\exists\, M \in \mathfrak{R} \ni \forall\, i \in \mathfrak{N}, \ \sum_{j=1}^{+\infty} |\, t_{ij}\,| \leqq M,$$

(2) $$\lim_{i \to +\infty} \sum_{j=1}^{+\infty} t_{ij} = 1,$$

(3) $$\forall\, j \in \mathfrak{N}, \ \lim_{i \to +\infty} t_{ij} = 0.$$

Let T be the Toeplitz matrix (t_{ij}), where $t_{ij} \equiv 1/i$ if $1 \leqq j \leqq i$ and $t_{ij} \equiv 0$ if $i < j$:

$$T = \begin{pmatrix} 1 & 0 & 0 & 0 & \cdots \\ \frac{1}{2} & \frac{1}{2} & 0 & 0 & \cdots \\ \frac{1}{3} & \frac{1}{3} & \frac{1}{3} & 0 & \cdots \\ \frac{1}{4} & \frac{1}{4} & \frac{1}{4} & \frac{1}{4} & \cdots \\ & & \cdots & & \end{pmatrix}.$$

The sequence $\{a_j\} = 1, -1, 1, -1, \cdots, (-1)^{n+1}, \cdots$, does not converge, but its transform by T,

$$\{b_i\} = 1, 0, \tfrac{1}{3}, 0, \tfrac{1}{5}, \cdots, \frac{1 + (-1)^{i+1}}{2i}, \cdots,$$

converges to 0.

More generally, if $\{a_n\}$ is *any* divergent sequence each of whose terms is either 1 or -1, there exists a Toeplitz matrix T that transforms $\{a_n\}$ into a convergent sequence. In fact, T can be defined so that $\{a_n\}$ is transformed into the sequence every term of which is 0. Such a matrix $T = (t_{ij})$ can be defined as follows: Let $\{n_i\}$ be a strictly increasing sequence of positive integers such that a_{n_i} and a_{n_i+1} have opposite signs for $i = 1, 2, \cdots$, and let

$$t_{ij} \equiv \begin{cases} \frac{1}{2} & \text{if } j = n_i \text{ or } j = n_i + 1, \\ 0 & \text{otherwise.} \end{cases}$$

Then T is a Toeplitz matrix that transforms $\{a_n\}$ into $0, 0, \cdots$.

[†] Named after the German mathematician Otto Toeplitz (1881–1940).

21. For a given Toeplitz matrix $T = (t_{ij})$, a sequence $\{a_j\}$ where for each j, $a_j = \pm 1$, such that the transform $\{b_i\}$ of $\{a_j\}$ by T diverges.

By reference to conditions (1)–(3) of Example 20, we choose two sequences $i_1 < i_2 < i_3 < \cdots$, $j_1 < j_2 < j_3 < \cdots$ as follows. Let i_1 be such that, in accordance with (2), if $i \geq i_1$,

$$\sum_{j=1}^{+\infty} t_{ij} = 1 + e_{1i}, \qquad |e_{1i}| < 0.05.$$

Then let j_1 be such that according to (1) and (2),

$$\sum_{j=1}^{j_1} t_{i_1 j} = 1 + d_1 \quad \text{and} \quad \sum_{j=j_i+1}^{+\infty} |t_{i_1 j}| < 0.05,$$

where $|d_1| < 0.1$.

Next choose $i_2 > i_1$ so that for $i \geq i_2$, according to (2) and (3),

$$\sum_{j=1}^{j_1} |t_{ij}| < (0.05) \text{ and } \sum_{j=1}^{+\infty} t_{ij} = 1 + e_{2i}, |e_{2i}| < (0.05)^2,$$

and then choose $j_2 > j_1$ so that according to (1) and (2),

$$\sum_{j=1}^{j_2} t_{i_2 j} = 1 + d_2 \quad \text{and} \quad \sum_{j=j_2+1}^{+\infty} |t_{i_2 j}| < (0.05)^2,$$

where $|d_2| < 2(0.05)^2$.

Having chosen $i_1 < i_2 < \cdots < i_k$ and $j_1 < j_2 < \cdots < j_k$, choose $i_{k+1} > i_k$ so that according to (2) and (3), for $i \geq i_{k+1}$,

$$\sum_{j=1}^{j_k} |t_{ij}| < (0.05)^k \text{ and } \sum_{j=1}^{+\infty} t_{ij} = 1 + e_{k+1,i}, |e_{k+1,i}| < (0.05)^{k+1},$$

and then choose $j_{k+1} > j_k$ so that according to (1) and (2),

$$\sum_{j=1}^{j_{k+1}} t_{i_{k+1} j} = 1 + d_{k+1}, \quad \text{and} \quad \sum_{j=j_{k+1}+1}^{+\infty} |t_{i_{k+1} j}| < (0.05)^{k+1},$$

where $|d_{k+1}| < 2(0.05)^{k+1}$.

Define the sequence $\{a_j\}$ by the prescription:

$$a_j \equiv \begin{cases} 1 \text{ for } 1 \leq j \leq j_1, j_2 < j \leq j_3, \cdots \\ -1 \text{ for } j_1 < j \leq j_2, j_3 < j \leq j_4, \cdots. \end{cases}$$

If k is odd and $k > 1$, then

$$b_{i_k} = \sum_{j=1}^{j_1} t_{i_k j} - \sum_{j=j_1+1}^{j_2} t_{i_k j} + \sum_{j=j_2+1}^{j_3} t_{i_k j} - \cdots$$

$$+ \sum_{j=j_{k-1}+1}^{j_k} t_{i_k j} + \sum_{j=j_k+1}^{\infty} t_{i_k j} a_j$$

$$= \sum_{j=1}^{j_1} t_{i_k j} - \left(\sum_{j=1}^{j_2} t_{i_k j} - \sum_{j=1}^{j_1} t_{i_k j} \right) + \left(\sum_{j=1}^{j_3} t_{i_k j} - \sum_{j=1}^{j_2} t_{i_k j} \right)$$

$$- \cdots + \left(\sum_{j=1}^{j_k} t_{i_k j} - \sum_{j=1}^{j_{k-1}} t_{i_k j} \right) + \sum_{j=j_k+1}^{+\infty} t_{i_k j} a_j.$$

All quantities $\sum_{j=1}^{j_r} t_{i_k j}$, except $\sum_{j=1}^{j_k} t_{i_k j} = 1 + d_k$, are less than $(0.05)^{k-1}$ in absolute value. Thus, since

$$\left| \sum_{j=j_k+1}^{+\infty} t_{i_k j} a_j \right| \leq \sum_{j=j_k+1}^{+\infty} | t_{i_k j} | < (0.05)^k,$$

we see that

$$b_{i_k} > 1 - 2(0.05)^{k-1} - 2(k - 1)(0.05)^{k-1} - (0.05)^k$$

$$= 1 - [2(k - 1) + 2.05](0.05)^{k-1}.$$

Methods of calculus show that the function of k defined on the last line above is an increasing function for $k \geq 1$, and therefore takes on its least value if $k \geq 3$ at the point where $k = 3$. Therefore, if k is an odd integer greater than 1:

$$b_{i_k} > 1 - [4 + 2.05](0.05)^2 > 0.9.$$

Similarly, for k even, it can be shown that $b_{i_k} < -0.7$. Therefore the sequence $\{b_i\}$ diverges.

Examples 20 and 21 show that although *some* Toeplitz matrices transform *some* sequences whose terms are all ± 1 into convergent sequences, *no* Toeplitz matrix transforms *all* such sequences into convergent sequences.

A refinement of the preceding technique permits the following conclusion: *If $\{T_m\}$ is a sequence of Toeplitz matrices, there is a sequence $\{a_n\}$ such that $| a_n | = 1$ for $n = 1, 2, \cdots$ and such that for each m the transform of $\{a_n\}$ by T_m is divergent.* Indeed, in the context preceding, let $\{i_k\}$ and $\{j_k\}$ be strictly increasing sequences of positive integers chosen so that the related properties hold as follows: We choose i_1 and j_1 so that the related properties hold for T_1; then i_2 and j_2 so that the related properties hold for both T_1 and T_2; etc. Let the sequence $\{a_n\}$ be defined as in Example 21. For any fixed m the sequence $\{a_n\}$ is transformed into a sequence $\{b_{mn}\}$ and, since the numbers i_k and j_k for $k > m$ constitute sequences valid for the counterexample technique applied to T_m, it follows that $\lim_{n \to +\infty} b_{mn}$ does not exist for any m.

22. A power series convergent at only one point. (Cf. Example 24.)

The series $\sum_{n=0}^{+\infty} n! \, x^n$ converges for $x = 0$ and diverges for $x \neq 0$.

23. A function whose Maclaurin series converges everywhere but represents the function at only one point.

The function

$$f(x) \equiv \begin{cases} e^{-1/x^2} & \text{if } x \neq 0, \\ 0 & \text{if } x = 0 \end{cases}$$

is infinitely differentiable, all of its derivatives at $x = 0$ being equal to 0 (cf. Example 10, Chapter 3). Therefore its Maclaurin series

$$\sum_{n=0}^{+\infty} \frac{f^{(n)}(0)}{n!} x^n = \sum_{n=0}^{+\infty} 0$$

converges for all x to the function that is identically zero, and hence represents (converges to) the given function f only at the single point $x = 0$.

24. A function whose Maclaurin series converges at only one point.

A function with this property is described in [10], p. 153. The function

$$f(x) = \sum_{n=0}^{+\infty} e^{-n} \cos n^2 x,$$

because of the factors e^{-n} present in all of the series obtained by successive term-by-term differentiation (which therefore all converge uniformly), is an infinitely differentiable function. Its Maclaurin series has only terms of even degree, and the absolute value of the term of degree $2k$ is

$$\sum_{n=0}^{+\infty} \frac{x^{2k} e^{-n} n^{4k}}{(2k)!} > \left(\frac{n^2 x}{2k}\right)^{2k} e^{-n}$$

for every $n = 0, 1, 2, \cdots$, and in particular for $n = 2k$. For this value of n and, in terms of any given nonzero x, with k any integer greater than $e/2x$, we have

$$\left(\frac{n^2 x}{2k}\right)^{2k} e^{-n} = \left(\frac{2kx}{e}\right)^{2k} > 1.$$

This means that for any nonzero x the Maclaurin series for f diverges.

The series $\sum_{n=0}^{+\infty} n!\, x^n$ was shown in Example 22 to be convergent at only one point, $x = 0$. It is natural to ask whether this series is the Maclaurin series for some function $f(x)$, since an affirmative answer would provide another example of a function of the type described in the present instance. We shall now show that it is indeed possible to produce an infinitely differentiable function $f(x)$ having the series given above as its Maclaurin series. To do this, let $\phi_{n0}(x)$ be defined as follows: For $n = 1, 2, \cdots$, let

$$\phi_{n0}(x) \equiv \begin{cases} ((n-1)!)^2 & \text{if } 0 \leq |x| \leq 2^{-n}/(n!)^2 \\ 0 & \text{if } |x| \geq 2^{-n+1}/(n!)^2 \end{cases}$$

where, by means of the type of "bridging functions" constructed for Example 12, Chapter 3, $\phi_{n0}(x)$ is made infinitely differentiable everywhere. Let $f_1(x) \equiv \phi_{10}(x)$, and for $n = 2, 3, \cdots$, let

$$\phi_{n1}(x) \equiv \int_0^x \phi_{n0}(t)\, dt,$$

$$\phi_{n2}(x) \equiv \int_0^x \phi_{n1}(t)\, dt,$$

$$\vdots$$

$$f_n(x) \equiv \phi_{n,n-1}(x) \equiv \int_0^x \phi_{n,n-2}(t)\, dt.$$

Thus $f_n'(x) = \phi_{n,n-2}(x)$, $f_n''(x) = \phi_{n,n-3}(x)$, $\cdots$, $f_n^{(n-1)}(x) = \phi_{n0}(x)$, $f_n^{(n)}(x) = \phi_{n0}'(x)$. For any x and $0 \leqq k \leqq n - 2$, $|f_n^{(k)}(x)| \leqq (2^{-n+1}/n^2)|x|^{n-k-2}/(n - k - 2)!$ since

$$|\phi_{n1}(x)| \leqq 2^{-n+1}/n^2,$$

$$|\phi_{n2}(x)| \leqq [2^{-n+1}/n^2] \cdot |x|,$$

$$\vdots$$

$$|\phi_{n,n-1}(x)| \leqq (2^{-n+1}/n^2) \cdot |x|^{n-2}/(n - 2)!.$$

The series $\sum_{n=1}^{+\infty} f_n^{(k)}(x)$, for each $k = 0, 1, 2, \cdots$, converges uniformly in every closed finite interval. Indeed, if $|x| \leqq K$,

$$\sum_{n=k+2}^{+\infty} |f_n^{(k)}(x)| \leqq \sum_{n=k+2}^{+\infty} \frac{K^{n-k-2}}{n^2 \cdot 2^{n-1} \cdot (n - k - 2)!},$$

and uniform convergence follows from the Weierstrass M-test (cf. [34], p. 445). Hence we see that

$$f(x) \equiv \sum_{n=1}^{+\infty} f_n(x)$$

is an infinitely differentiable function such that for $k = 0, 1, 2, \cdots$,

$$f^{(k)}(x) = \sum_{n=1}^{+\infty} f_n^{(k)}(x).$$

For $k \geqq n \geqq 1$, $f_n^{(k)}(0) = \phi_{n0}^{(k-n+1)}(0) = 0$. For $n \geqq 1$ and $k = n - 1$, $f_n^{(k)}(0) = \phi_{n0}(0) = ((n - 1)!)^2$. For $0 \leqq k < n - 1$, $f_n^{(k)}(0) = 0$. Thus the Maclaurin series for $f(x)$ is $\sum_{n=0}^{+\infty} n! \, x^n$.

25. A convergent trigonometric series that is not a Fourier series.

We shall present two examples, one in case the integration involved is that of Riemann, and one in case the integration is that of Lebesgue.

The series $\sum_{n=1}^{+\infty} \dfrac{\sin nx}{n^\alpha}$, where $0 < \alpha \leqq \frac{1}{2}$, converges for every real number x, as can be seen (cf. [34], p. 533) by an application of a convergence test due to N. H. Abel (Norwegian, 1802–1829). However, this series cannot be the Fourier series of any Riemann-integrable function $f(x)$ since, by Bessel's inequality (cf. [34], p. 532), for $n = 1, 2, \cdots$,

$$\frac{1}{1^{2\alpha}} + \frac{1}{2^{2\alpha}} + \cdots + \frac{1}{n^{2\alpha}} \leq \frac{1}{\pi} \int_{-\pi}^{\pi} [f(x)]^2 \, dx.$$

Since $f(x)$ is Riemann-integrable so is $[f(x)]^2$, and the right-hand side of the preceding inequality is finite, whereas if $\alpha \leq \frac{1}{2}$ the left-hand side is unbounded as $n \to +\infty$. (Contradiction.)

The series $\sum\limits_{n=2}^{+\infty} \dfrac{\sin nx}{\ln n}$ also converges for every real number x. Let

$$f(x) \equiv \sum_{n=2}^{+\infty} \frac{\sin nx}{\ln n}.$$

If $f(x)$ is Lebesgue-integrable, then the function

$$F(x) \equiv \int_0^x f(t) \, dt$$

is both periodic and absolutely continuous. Since $f(x)$ is an odd function $(f(x) = -f(-x))$, we see that $F(x)$ is an even function $(F(x) = F(-x))$ and thus the Fourier series for $F(x)$ is of the form

$$\sum_{n=0}^{+\infty} a_n \cos nx,$$

where $a_0 = \dfrac{1}{\pi} \int_0^\pi F(x) \, dx$, and for $n \geq 2$,

$$a_n = \frac{2}{\pi} \int_0^\pi F(x) \cos nx \, dx$$

$$= \frac{2}{\pi} F(x) \frac{\sin nx}{n} \Big|_0^\pi - \frac{2}{\pi} \int_0^\pi F'(x) \frac{\sin nx}{n} \, dx$$

$$= -\frac{2}{\pi} \int_0^\pi f(x) \frac{\sin nx}{n} \, dx = -\frac{1}{n \ln n}.$$

($F'(x)$ exists and is equal to $f(x)$ almost everywhere.) Since $F(x)$ is of bounded variation, its Fourier series converges at every point, and in particular at $x = 0$, from which we infer that $\sum_{n=2}^{+\infty} a_n$ converges. But since $a_n = -1/(n \ln n)$, and $\sum_{n=2}^{+\infty} (-1/(n \ln n))$ diverges, we have a contradiction to the assumption that $f(x)$ is Lebesgue-integrable.

Finally, to show that the trigonometric series $\sum_{n=2}^{+\infty} \dfrac{\sin nx}{\ln n}$ is not the Fourier series of *any* Lebesgue-integrable function we need the following theorem (for a proof see [52] and [53]): *If a trigonometric series is the Fourier series of a Lebesgue-integrable function g, and if it converges almost everywhere to a function f, then $f(x) = g(x)$ almost everywhere and therefore f is Lebesgue-integrable and the given trigonometric series is the Fourier series of f.*

26. An infinitely differentiable function $f(x)$ such that

$$\lim_{|x| \to +\infty} f(x) = 0$$

and that is not the Fourier transform of any Lebesgue-integrable function.

Let $\{c_n\}$, $n = 0, \pm 1, \pm 2, \cdots$, be a doubly infinite sequence such that

$$\sum_{n=-\infty}^{+\infty} c_n e^{inx}$$

converges for all x but does not represent a function Lebesgue-integrable on $[-\pi, \pi]$ (cf. Example 25). We shall show that if $h(x)$ is any infinitely differentiable function that vanishes outside $[-\frac{1}{2}, \frac{1}{2}]$, if $h(0) = 2\pi$, and if

$$f(x) \equiv \sum_{n=-\infty}^{+\infty} c_n h(x - n),$$

then $f(x)$ is a function of the required kind.

Since $h(x)$ vanishes outside $[-\frac{1}{2}, \frac{1}{2}]$, the series that defines $f(x)$ has only finitely many nonzero terms for any fixed x. Hence, the series converges for all x and represents some function $f(x)$. By the same argument, the series converges uniformly in every finite interval as does the series arising by termwise differentiation any finite number of times, and furthermore

$$f^{(k)}(x) = \sum_{n=-\infty}^{+\infty} c_n h^{(k)}(x - n),$$

$k = 0, 1, 2, \cdots$.

If $F(t)$ is Lebesgue-integrable and satisfies

$$\int_{-\infty}^{+\infty} F(t)e^{-itx} \, dt = f(x),$$

let

$$g(t) \equiv \sum_{m=-\infty}^{+\infty} F(t + 2\pi m).$$

Since $F(t)$ is Lebesgue integrable, $g(t)$ is defined for almost every t, $g(t + 2\pi) = g(t)$, and

$$\int_{-\pi}^{\pi} |g(t)| \, dt \leq \sum_{m=-\infty}^{+\infty} \int_{-\pi}^{\pi} |F(t + 2\pi m)| \, dt = \int_{-\infty}^{+\infty} |F(t)| \, dt < +\infty.$$

(For references on the preceding facts and the following equality, see [29], pp. 130–132, pp. 152–153.) We compute:

$$\frac{1}{2\pi} \int_{-\pi}^{\pi} g(t)e^{-int} \, dt = \frac{1}{2\pi} \sum_{m=-\infty}^{+\infty} \int_{-\pi}^{\pi} F(t + 2\pi m)e^{-int} \, dt$$

$$= \frac{1}{2\pi} \sum_{m=-\infty}^{+\infty} \int_{-\pi}^{\pi} F(t + 2\pi m)e^{-in(t+2\pi m)} \, dt$$

$$= \frac{1}{2\pi} \int_{-\infty}^{+\infty} F(t)e^{-int} \, dt = \frac{1}{2\pi} f(n)$$

$$= \frac{1}{2\pi} \sum_{k=-\infty}^{+\infty} c_k h(n - k) = \frac{c_n}{2\pi} h(0) = c_n.$$

In other words, the c_n are Fourier coefficients of the Lebesgue-integrable function $g(t)$. This contradiction establishes the fact that $F(x)$ is not the Fourier transform of a Lebesgue-integrable function.

27. For an arbitrary countable set $E \subset [-\pi, \pi]$, a continuous function whose Fourier series diverges at each point of E and converges at each point of $[-\pi, \pi] \setminus E$.

The idea behind this example goes back to Fejér and Lebesgue. An exposition of the details is given in [52], pp. 167–173, where references to the original papers are also to be found.

28. A (Lebesgue-) integrable function on $[-\pi, \pi]$ whose Fourier series diverges everywhere.

This example is due to A. Kolmogorov. Details are given in [52], pp. 175–179, together with references.

29. A sequence $\{a_n\}$ of rational numbers such that for every function f continuous on $[0, 1]$ and vanishing at 0 ($f(0) = 0$) there exists a strictly increasing sequence $\{n_\nu\}$ of positive integers such that, with $n_0 \equiv 0$:

$$f(x) = \sum_{\nu=0}^{+\infty} \left(\sum_{n=n_\nu+1}^{n_{\nu+1}} a_n x^n \right),$$

the convergence being uniform on $[0, 1]$.

We make a preliminary observation:

For any positive integer m, the set of all polynomials with rational coefficients and involving only powers x^n such that $n \geqq m$ is dense in the space $C_0([0, 1])$ of all functions continuous on $[0, 1]$ and vanishing at 0, in the "uniform topology" given by the distance and norm formula:

$$\rho(f, g) = \| f - g \| \equiv \max \{ | f(x) - g(x) | \mid 0 \leqq x \leqq 1 \}.$$

This follows from the Stone-Weierstrass theorem (cf. [42], p. 288, and [29], pp. 9–10).

Now let $\{f_n\}$ be an enumeration of a countable dense set of functions in $C_0([0, 1])$. For example, $\{f_n\}$ might be a sequence consisting of all polynomials with rational coefficients and zero constant term. Let P_1 be such a polynomial for which $\rho(f_1, P_1) = \| f_1 - P_1 \| < 1$. Let P_2 be such a polynomial for which, among the terms with nonzero coefficients, the least exponent exceeds the degree of P_1, and for which $\rho(f_2 - P_1, P_2) = \| f_2 - (P_1 + P_2) \| < \frac{1}{2}$. Having defined polynomials $P_1, P_2, \cdots, P_n$ in $C_0([0, 1])$ such that their coefficients are rational and such that:

(a) the "least exponent of P_{k+1}" > deg P_k, $k = 1, 2, \cdots, n - 1$,

(b) $\left\| f_k - \sum_{i=1}^{k} P_i \right\| < \frac{1}{k}$, $k = 1, 2, \cdots, n$,

choose a polynomial P_{n+1} in $C_0([0, 1])$ with rational coefficients and such that

(a') the "least exponent of P_{n+1}" > deg P_n,

(b') $\left\| f_{n+1} - \sum_{i=1}^{n+1} P_i \right\| < \dfrac{1}{n+1}$.

Let $m_j \equiv$ "least exponent of P_j," and let $M_j \equiv \deg P_j$ for every $j \in \mathfrak{N}$. Then $m_j \leqq M_j < m_{j+1}$, for $j \in \mathfrak{N}$. The sequence $\{a_n\}$ is now defined as follows: $a_1 = a_2 = \cdots = a_{m_1 - 1} \equiv 0$, $a_{m_1} \equiv$ coefficient of x^{m_1} in P_1, $a_{m_1+1} =$ coefficient of x^{m_1+1} in P_1, $\cdots$, $a_{M_1} \equiv$ coefficient of x^{M_1} in P_1. In general for $M_j < n < m_{j+1}$ let $a_n \equiv 0$; for $m_{j+1} \leqq n \leqq M_{j+1}$, let $a_n \equiv$ coefficient of x^n in P_j. If $f \in C_0([0, 1])$, let $0 < k_1 < k_2 < \cdots$ be such that $\| f - f_{k_\mu} \| < 1/\mu$ for every $\mu \in \mathfrak{N}$. Then

$$\left\| f - \sum_{i=1}^{k_\mu} P_i \right\| \leqq \| f - f_{k_\mu} \| + \left\| f_{k_\mu} - \sum_{i=1}^{k_\mu} P_i \right\| < \frac{1}{\mu} + \frac{1}{k_\mu} \leqq \frac{2}{\mu}.$$

Hence if $n_0 \equiv 0$ and $n_\nu \equiv M_{k_\nu}$ for $\nu \in \mathfrak{N}$, then

$$f(x) = \sum_{\nu=0}^{+\infty} \left(\sum_{n=n_\nu+1}^{n_{\nu+1}} a_n x^n \right),$$

where the (grouped) series on the right converges uniformly in $[0, 1]$.

This startling result is due to W. Sierpinski. Its close similarity to Example 5, last paragraph, should be noted. In this latter case, a single series of numbers is obtained, as the result of a rearrangement, having the property that corresponding to an arbitrary real number x there exists a subsequence of partial sums — and hence a method for introducing parentheses in the series — that gives convergence to x. In the present case there is a single power series having the property that corresponding to an arbitrary member of $C_0([0, 1])$ there exists a subsequence of partial sums — and hence a method of introducing parentheses in the series — that gives uniform convergence to f.

Chapter 7
Uniform Convergence

Introduction

The examples of this chapter deal with uniform convergence — and convergence that is *not* uniform — of sequences of functions on certain sets. The basic definitions and theorems will be assumed to be known (cf. [34], pp. 441–462, [36], pp. 270–292).

1. A sequence of everywhere discontinuous functions converging uniformly to an everywhere continuous function.

$$f_n(x) \equiv \begin{cases} 1/n & \text{if } x \text{ is rational,} \\ 0 & \text{if } x \text{ is irrational.} \end{cases}$$

Clearly, $\lim_{n \to +\infty} f_n(x) = 0$ uniformly for $-\infty < x < +\infty$.

This simple example serves to illustrate the following general principle: *Uniform convergence preserves good behavior, not bad behavior.* This same principle will be illustrated repeatedly in future examples.

2. A sequence of infinitely differentiable functions converging uniformly to zero, the sequence of whose derivatives diverges everywhere.

If $f_n(x) \equiv (\sin nx)/\sqrt{n}$, then since $|f_n(x)| \leq 1/\sqrt{n}$ this sequence converges uniformly to 0. To see that the sequence $\{f_n'(x)\}$ converges nowhere, let x be fixed and consider

$$b_n = f_n'(x) = \sqrt{n} \cos nx.$$

If $x = 0$, $b_n = \sqrt{n} \to +\infty$ as $n \to +\infty$. We shall show that for any $x \neq 0$ the sequence $\{b_n\}$ is unbounded, and hence diverges, by showing that there are arbitrarily large values of n such that $|\cos nx| \geqq \frac{1}{2}$. Indeed, for any positive integer m such that $|\cos mx| < \frac{1}{2}$,

$$|\cos 2mx| = |2\cos^2 mx - 1| = 1 - 2\cos^2 mx > \tfrac{1}{2},$$

so that there exists an $n > m$ such that $|\cos nx| > \frac{1}{2}$.

3. A nonuniform limit of bounded functions that is not bounded.

Each function

$$f_n(x) \equiv \begin{cases} \min\left(n, \dfrac{1}{x}\right) & \text{if } \ 0 < x \leqq 1, \\[2mm] 0 & \text{if } \ x = 0 \end{cases}$$

is bounded on the closed interval $[0, 1]$, but the limit function $f(x)$, equal to $1/x$ if $0 < x \leqq 1$ and equal to 0 if $x = 0$, is unbounded there.

Let it be noted that for this example to exist, the limit *cannot* be uniform.

4. A nonuniform limit of continuous functions that is not continuous.

A trivial example is given by

$$f_n(x) \equiv \begin{cases} \min\,(1, nx) & \text{if } \ x \geqq 0, \\ \max\,(-1, nx) & \text{if } \ x < 0, \end{cases}$$

whose limit is the signum function (Example 3, Chapter 3), which is discontinuous at $x = 0$.

A more interesting example is given by use of the function f (cf. Example 15, Chapter 2) defined:

$$f(x) \equiv \begin{cases} \dfrac{1}{q} & \text{if } \ x = \dfrac{p}{q} \text{ in lowest terms, where } p \text{ and } q \text{ are integers} \\ & \qquad\qquad \text{and } q > 0. \\[2mm] 0 & \text{if } \ x \text{ is irrational.} \end{cases}$$

For an arbitrary positive integer n, define $f_n(x)$ as follows: According to each point $\left(\dfrac{p}{q}, \dfrac{1}{q}\right)$, where $1 \leqq q < n$, $0 \leqq p \leqq q$, in each interval of the form $\left(\dfrac{p}{q} - \dfrac{1}{2n^2}, \dfrac{p}{q}\right)$ define

$$f_n(x) \equiv \min\left(\frac{1}{n}, \frac{1}{q} + 2n^2\left(x - \frac{p}{q}\right)\right);$$

in each interval of the form $\left(\dfrac{p}{q}, \dfrac{p}{q} + \dfrac{1}{2n^2}\right)$ define

$$f_n(x) \equiv \max\left(\frac{1}{n}, \frac{1}{q} - 2n^2\left(x - \frac{p}{q}\right)\right);$$

and at every point x of $[0, 1]$ at which $f_n(x)$ has not already been defined, let $f_n(x) \equiv 1/n$. Outside $[0, 1]$ $f_n(x)$ is defined so as to be periodic with period one. The graph of $f_n(x)$, then, consists of an infinite polygonal arc made up of segments that either lie along the horizontal line $y = 1/n$ or rise with slope $\pm 2n^2$ to the isolated points of the graph of f. (Cf. Fig. 2.) As n increases, these "spikes" sharpen, and the base approaches the x axis. As a consequence, for each $x \in \Re$ and $n = 1, 2, \cdots$,

$$f_n(x) \geqq f_{n+1}(x),$$

and

$$\lim_{n \to +\infty} f_n(x) = f(x),$$

as defined above. Each function f_n is everywhere continuous, but the limit function f is discontinuous on the dense set Q of rational numbers. (Cf. Example 24, Chapter 2.)

5. A nonuniform limit of Riemann-integrable functions that is not Riemann-integrable. (Cf. Example 33, Chapter 8.)

Each function g_n, defined for Example 24, Chapter 2, when restricted to the closed interval $[0, 1]$ is Riemann-integrable there, since it is bounded there and has only a finite number of points of discontinuity. The sequence $\{g_n\}$ is an increasing sequence ($g_n(x) \leqq g_{n+1}(x)$ for each x and $n = 1, 2, \cdots$) converging to the function f of Example 1, Chapter 4, that is equal to 1 on $Q \cap [0, 1]$ and equal to 0 on $[0, 1] \setminus Q$.

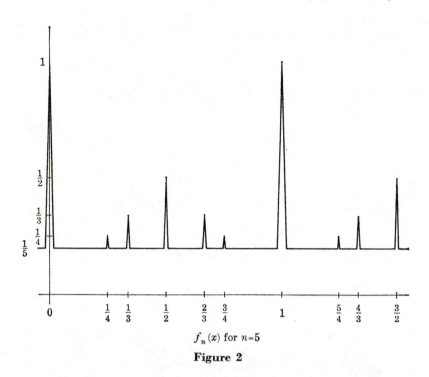

$f_n(x)$ for $n=5$

Figure 2

6. A sequence of functions for which the limit of the integrals is not equal to the integral of the limit.

Let

$$f_n(x) \equiv \begin{cases} 2n^2x & \text{if} \quad 0 \leqq x \leqq \dfrac{1}{2n}, \\[2mm] n - 2n^2\left(x - \dfrac{1}{2n}\right) & \text{if} \quad \dfrac{1}{2n} \leqq x \leqq \dfrac{1}{n}, \\[2mm] 0 & \text{if} \quad \dfrac{1}{n} \leqq x \leqq 1. \end{cases}$$

Then

$$\lim_{n \to +\infty} \int_0^1 f_n(x)\, dx = \lim_{n \to +\infty} \frac{1}{2} = \frac{1}{2},$$

but

$$\int_0^1 \lim_{n \to +\infty} f_n(x)\, dx = \int_0^1 0\, dx = 0.$$

79

Another example is the sequence $\{f_n(x)\}$ where $f_n(x) \equiv nxe^{-nx}$, $0 \leq x \leq 1$.

A more extreme case is given by

$$f_n(x) \equiv \begin{cases} 2n^3x & \text{if } 0 \leq x \leq \dfrac{1}{2n}, \\[2mm] n^2 - 2n^3\left(x - \dfrac{1}{2n}\right) & \text{if } \dfrac{1}{2n} \leq x \leq \dfrac{1}{n}, \\[2mm] 0 & \text{if } \dfrac{1}{n} \leq x \leq 1, \end{cases}$$

in which case, for any $b \in (0, 1]$

$$\lim_{n \to +\infty} \int_0^b f_n(x)\, dx = \lim_{n \to +\infty} \frac{n}{2} = +\infty,$$

while

$$\int_0^b \lim_{n \to +\infty} f_n(x)\, dx = \int_0^b 0\, dx = 0.$$

7. A sequence of functions for which the limit of the derivatives is not equal to the derivative of the limit.

If $f_n(x) \equiv x/(1 + n^2x^2)$ for $-1 \leq x \leq 1$ and $n = 1, 2, \cdots$, then $f(x) \equiv \lim_{n \to +\infty} f_n(x)$ exists and is equal to 0 for all $x \in [-1, 1]$ (and this convergence is uniform since the maximum and minimum values of $f_n(x)$ on $[-1, 1]$ are $\pm 1/2n$). The derivative of the limit is identically equal to 0. However, the limit of the derivatives is

$$\lim_{n \to +\infty} f_n'(x) = \lim_{n \to +\infty} \frac{1 - n^2x^2}{(1 + n^2x^2)^2} = \begin{cases} 1 & \text{if } x = 0, \\ 0 & \text{if } 0 < |x| \leq 1. \end{cases}$$

8. Convergence that is uniform on every closed subinterval but not uniform on the total interval.

Let $f_n(x) \equiv x^n$ on the open interval $(0, 1)$.

9. A sequence $\{f_n\}$ converging uniformly to zero on $[0, +\infty)$ but such that $\int_0^{+\infty} f_n(x)\, dx \nrightarrow 0$.

Let $f_n(x) \equiv \begin{cases} 1/n & \text{if } 0 \leq x \leq n, \\ 0 & \text{if } x > n. \end{cases}$

Then f_n converges uniformly to 0 on $[0, +\infty)$, but

$$\int_0^{+\infty} f_n(x)\, dx = 1 \to 1.$$

A more extreme case is given by

$$f_n(x) \equiv \begin{cases} 1/n & \text{if } 0 \leqq x \leqq n^2, \\ 0 & \text{if } x > n^2. \end{cases}$$

Then $\int_0^{+\infty} f_n(x)\, dx = n \to +\infty$.

10. A series that converges nonuniformly and whose general term approaches zero uniformly.

The series $\sum_{n=1}^{+\infty} x^n/n$ on the half-open interval $[0, 1)$ has these properties. Since the general term is dominated by $1/n$ on $[0, 1)$ its uniform convergence to zero there follows immediately. The convergence of the series follows from its domination by the series $\sum x^n$, which converges on $[0, 1)$. The nonuniformity of this convergence is a consequence of the fact that the partial sums are not uniformly bounded (the harmonic series diverges; cf. [34], p. 447, Exs. 31, 32).

11. A sequence converging nonuniformly and possessing a uniformly convergent subsequence.

On the real number system $\mathfrak{R}$, let

$$f_n(x) \equiv \begin{cases} \dfrac{x}{n} & \text{if } n \text{ is odd.} \\[2mm] \dfrac{1}{n} & \text{if } n \text{ is even.} \end{cases}$$

The convergence to zero is nonuniform, but the convergence of the subsequence $\{f_{2n}(x)\} = \{1/2n\}$ is uniform.

12. Nonuniformly convergent sequences satisfying any three of the four conditions of Dini's theorem.

Dini's theorem states that if $\{f_n\}$ is a sequence of functions defined on a set A and converging on A to a function f, and if

(i) f_n is continuous on A, $n = 1, 2, \cdots$,

(*ii*) f is continuous on A,

(*iii*) the convergence is monotonic,

(*iv*) A is compact,

then the convergence is uniform.

No three of these conditions imply uniform convergence. In other words, no one of the four hypotheses can be omitted. The following four examples demonstrate this fact.

$$(i): f_n(x) \equiv \begin{cases} 0 & \text{if} \quad x = 0 \quad \text{or} \quad \frac{1}{n} \leqq x \leqq 1, \\ 1 & \text{if} \quad 0 < x < \frac{1}{n}. \end{cases}$$

Then $\{f_n(x)\}$ is a decreasing sequence for each x, converging nonuniformly to the continuous function 0 on the compact set $[0, 1]$.

(*ii*): The sequence $\{x^n\}$ converges decreasingly and nonuniformly to the discontinuous function

$$f(x) \equiv \begin{cases} 0 & \text{if} \quad 0 \leqq x < 1, \\ 1 & \text{if} \quad x = 1 \end{cases}$$

on the compact set $[0, 1]$.

(*iii*): Example 6.

(*iv*): The sequence $\{x^n\}$ on $[0, 1)$.

Chapter 8
Sets and Measure on the Real Axis

Introduction

Unless a specific statement to the contrary is made, all sets considered in this chapter should be assumed to be subsets of $\Re$, the real number system. A σ-**ring**, or **sigma-ring**, is a nonempty class A of sets that is closed under the operations of countable unions and set differences ($A_1, A_2, \cdots \in A \Rightarrow \bigcup_{n=1}^{+\infty} A_n \in A$ and $A_1 \setminus A_2 \in A$). If A is any nonempty class of sets, the σ-ring **generated** by A is the intersection of all σ-rings containing A (there is always at least *one* σ-ring containing A, the class of *all* subsets of $\Re$, so that the generated σ-ring always exists). It is natural to think of the σ-ring generated by A as the *smallest* σ-ring containing A. The σ-ring generated by the class C of all compact subsets of $\Re$ is called the class B of **Borel sets** (that is, a set is a Borel set iff it is a member of the σ-ring generated by C).

If A is any subset of $\Re$ and x any real number, the **translate** of A by x is defined and denoted:

$$x + A \equiv \{y \mid y = x + a, a \in A\} = \{x + a \mid a \in A\}.$$

A class A of sets is **closed under translations** iff

$$A \in A, x \in \Re \Rightarrow x + A \in A.$$

If S is a σ-ring of subsets of a space X, a set-function ρ with domain S is **nonnegative extended-real-valued** iff its values, for sets $S \in S$, satisfy the inequalities $0 \leq \rho(S) \leq +\infty$. A nonnegative extended-real-valued set-function ρ on a σ-ring S is a **measure** on S

iff $\rho(\emptyset) = 0$ and ρ is **countably additive** on S:

$$S_1, S_2, \cdots \in S, S_m \cap S_n = \emptyset \text{ for } m \neq n \Rightarrow$$

$$\rho\left(\bigcup_{n=1}^{+\infty} S_n\right) = \sum_{n=1}^{+\infty} \rho(S_n).$$

If ρ is a measure on a σ-ring S of subsets of a space X, and if $X \in S$, then the ordered pair (X, S) is called a **measure space** and ρ is called a **measure** on the measure space (X, S). If the class of sets S is understood from context, the single letter X may also be used to indicate a measure space. If ρ and σ are two measures on the same measure space (X, S), ρ is **absolutely continuous** with respect to σ (written $\rho \ll \sigma$) iff

$$A \in S, \sigma(A) = 0 \Rightarrow \rho(A) = 0.$$

For any measure ρ on a measure space (X, S), a **null-set** for ρ is any subset of a member A of S of measure zero: $\rho(A) = 0$. A measure ρ on (X, S) is **complete** iff every null-set for ρ is a member of S.

Borel measure is the uniquely determined measure μ on the measure space $(\mathfrak{R}, B)$ that assigns to every bounded closed interval its length:

$$\mu([a, b]) = b - a \qquad \text{if} \quad a \leqq b.$$

The class $\tilde{B}$ of Lebesgue-measurable sets is the σ-ring generated by the union of B and the class of all null-sets of Borel measure on B. Lebesgue measure is the uniquely determined complete measure on $\tilde{B}$ whose contraction to B is Borel measure; that is, Lebesgue measure is the *completion*, or *complete extension*, to $\tilde{B}$ of Borel measure on B.

Since the length of a compact interval $[a, b]$ is invariant under translations, the σ-rings B and $\tilde{B}$ are closed under translations, and both Borel and Lebesgue measure are **translation-invariant**:

$$A \in B, x \in \mathfrak{R} \Rightarrow x + A \in B, \mu(x + A) = \mu(A),$$

$$A \in \tilde{B}, x \in \mathfrak{R} \Rightarrow x + A \in \tilde{B}, \mu(x + A) = \mu(A).$$

For any set $E \subset \mathfrak{R}$, *Lebesgue inner* and *outer measure* are defined and denoted:

inner measure of $E = \mu_*(E) \equiv \sup \{\mu(A) \mid A \subset E, A \in \tilde{B}\}$,

outer measure of $E = \mu^*(E) \equiv \inf \{\mu(A) \mid A \supset E, A \in \tilde{B}\}$.

These are also equal to

$$\mu_*(E) = \sup \{\mu(A) \mid A \subset E, A \in \boldsymbol{B}\}$$
$$= \sup\{\mu(A) \mid A \subset E, A \text{ compact}\},$$
$$\mu^*(E) = \inf \{\mu(A) \mid A \supset E, A \in \boldsymbol{B}\}$$
$$= \inf \{\mu(A) \mid A \supset E, A \text{ open}\}.$$

For proofs of the preceding facts, and for further discussion, see [16], [18], [30], and [32].

We shall occasionally refer to the axiom of choice, or such variants as the well-ordering theorem or Zorn's lemma. These are sometimes classed under the title of *The Maximality Principle*. The reader is referred to [16], [30], and [46].

It will be assumed that the reader is already familiar with the concept of *equivalence relations* and *equivalence classes*. These topics are treated in references [16] and [22].

1. A perfect nowhere dense set.

A **perfect set** is a closed set every point of which is a limit point of the set. A fundamental fact concerning perfect sets is that every nonempty perfect set A of real numbers — or, more generally, any nonempty perfect set in a complete separable metric space — is uncountable; in fact, A has the cardinality $\mathfrak{c}$ of $\mathfrak{R}$ (there exists a one-to-one correspondence with domain $\mathfrak{R}$ and range A). (For a proof and discussion, see [20], pp. 129–138.)

A **nowhere dense set** is a set A whose closure $\bar{A}$ has no interior points: $I(\bar{A}) = \emptyset$. Clearly, a set is nowhere dense iff its closure is nowhere dense, and any subset of a nowhere dense set is nowhere dense. A less obvious fact is that the union of any finite collection of nowhere dense sets is nowhere dense. Proof by induction follows from the special case: *If A and B are closed and nowhere dense, then $A \cup B$ is nowhere dense.* (If U is a nonempty open subset of $A \cup B$, then $U \setminus B$ is a nonempty open subset of A.)

A celebrated example of a perfect nowhere dense set was given by G. Cantor (German, 1845–1918), and is known as **the Cantor set**. This set C is obtained from the closed unit interval $[0, 1]$ by a sequence of deletions of open intervals known as "middle thirds," as follows: First delete all points x between $\frac{1}{3}$ and $\frac{2}{3}$. Then remove the middle thirds of the two closed intervals [0, $\frac{1}{3}$] and [$\frac{2}{3}$, 1] remaining: ($\frac{1}{9}$, $\frac{2}{9}$) and ($\frac{7}{9}$, $\frac{8}{9}$). Then remove the middle thirds of the four closed intervals [0, $\frac{1}{9}$], [$\frac{2}{9}$, $\frac{1}{3}$], [$\frac{2}{3}$, $\frac{7}{9}$], and [$\frac{8}{9}$, 1] remaining: ($\frac{1}{27}$, $\frac{2}{27}$), ($\frac{7}{27}$, $\frac{8}{27}$), ($\frac{19}{27}$, $\frac{20}{27}$), and ($\frac{25}{27}$, $\frac{26}{27}$). This process is permitted to continue indefinitely, with the result that the total set of points removed from [0, 1] is the union of a sequence of open intervals and hence is an open set. The set C is defined to be the closed set remaining. Since every point of C is approached arbitrarily closely by endpoints of intervals removed (these endpoints all belong to C), C is perfect. Since there is no open interval within [0, 1] that has no points in common with at least one of the open intervals whose points are deleted at some stage, the (closed) set C is nowhere dense.

The Cantor set C can be defined in terms of the ternary (base three) system of numeration. A point $x \in C$ iff x can be represented by means of a ternary expansion using *only* the digits 0 and 2. For example, 0.022222··· and 0.200000··· are the endpoints of the first interval removed, or $\frac{1}{3}$ and $\frac{2}{3}$, respectively, in decimal notation. For a discussion of this description of C, cf. [18] and [32]; also cf. Example 2, below.

2. An uncountable set of measure zero.

The Cantor set C of Example 1 is uncountable since it is a nonempty perfect set, and it has measure zero since the set of points deleted from the closed interval [0, 1] has measure

$$\frac{1}{3} + \frac{1}{3}\cdot\frac{2}{3} + \frac{1}{3}\cdot\frac{2}{3}\cdot\frac{2}{3} + \cdots = \frac{\frac{1}{3}}{1 - \frac{2}{3}} = 1.$$

The ternary expansions of the points of the Cantor set can be used to show that C has the cardinality $\mathfrak{c}$ of the real number system $\mathfrak{R}$. (This method is independent of the one cited above that is based on the properties of perfect sets.) In the first place, the points of C are in one-to-one correspondence with the ternary expansions using only the digits 0 and 2, and therefore (divide by 2) with the *binary* expansions using the digits 0 and 1. On the other hand, the *nonterminating* binary expansions are in one-to-one correspondence with the points of the half-open interval (0, 1], and hence with the real numbers. This much shows that the set of *all* binary expansions — and therefore C — is uncountable, with cardinality *at least* $\mathfrak{c}$. To see that the cardinality is actually *equal* to $\mathfrak{c}$ we need only observe that the set of *terminating* binary expansions is countable (or, even more simply, that $C \subset \Re$). For further discussion of the mapping just described, see Example 14, below.

3. A set of measure zero whose difference set contains a neighborhood of the origin.

If A is any nonempty set, its **difference set,** $D(A)$ is the set of all differences between members of A:

$$D(A) \equiv \{x - y \mid x \in A, y \in A\}.$$

A fact of some importance in measure theory is that whenever A is a measurable set of positive measure, the origin is an interior point of the difference set A (cf. [18], p. 68). The Cantor set C of Example 1 is an example of a set of measure *zero* that has this same property. In fact, the difference set of C is the entire closed interval $[-1, 1]$:

$$D(C) = [-1, 1].$$

The simplest way to see this is to consider the product set $C \times C$, and to show that for any number α such that $-1 \leq \alpha \leq 1$, the line $y = x + \alpha$ meets the set $C \times C$ in at least one point. (Cf. [10], p. 110, where references are given.) Since C is obtained by a sequence of removals of "middle thirds," the set $C \times C$ can be thought of as the intersection of a countable family of closed sets $C_1, C_2, \cdots$, each of the sets C_n being a union of "corner squares" as follows (cf. Fig. 3): The set C_1 consists of four $\frac{1}{3}$ by $\frac{1}{3}$ closed squares located in the corners of the total square $[0, 1] \times [0, 1] : [0, \frac{1}{3}] \times [0, \frac{1}{3}]$, $[0, \frac{1}{3}] \times [\frac{2}{3}, 1], [\frac{2}{3}, 1] \times [0, \frac{1}{3}]$, and $[\frac{2}{3}, 1] \times [\frac{2}{3}, 1]$; the set C_2 consists of sixteen $\frac{1}{9}$ by $\frac{1}{9}$ closed squares located by fours in the corners of the four squares of C_1; the set C_3 consists of sixty-four $\frac{1}{27}$ by $\frac{1}{27}$ squares located by fours in the corners of the sixteen squares of C_2; etc. For any given $\alpha \in [-1, 1]$, the line $y = x + \alpha$ meets at least one of the four squares of C_1; choose such a square and denote it S_1. This line must also meet at least one of the four squares of C_2 that lie within S_1; choose such a square and denote it S_2. If this process is continued, a sequence of closed squares $\{S_n\}$ is obtained such that

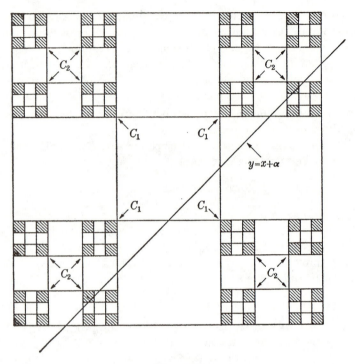

Figure 3

$S_{n+1} \subset S_n$ for $n = 1, 2, \cdots$. Since the side of S_n is 3^{-n}, there is exactly one point (x, y) that belongs to *every* square of the sequence $\{S_n\}$ (cf. [34], p. 201, Ex. 30). The point (x, y) must therefore belong to $C \times C$, and since this point must also lie on the line $y = x + \alpha$, we have the desired members x and y of C whose difference is the prescribed number α.

4. Perfect nowhere dense sets of positive measure.

The process used to obtain the Cantor set C of Example 1 can be modified to construct a useful family of perfect nowhere dense sets. Each of these sets, to be called **a Cantor set,** is the set of points remaining in [0, 1] after a sequence of deletions has taken place as follows: If α is an arbitrary positive number less than 1, first delete from $[0, 1]$ all points of the open interval $(\frac{1}{2} - \frac{1}{4}\alpha, \frac{1}{2} + \frac{1}{4}\alpha)$, of length $\frac{1}{2}\alpha$ and midpoint $\frac{1}{2}$. From the two remaining closed intervals $[0, \frac{1}{2} - \frac{1}{4}\alpha]$ and $[\frac{1}{2} + \frac{1}{4}\alpha, 1]$, each of length $\frac{1}{2}(1 - \frac{1}{2}\alpha)$, remove the middle open intervals each of length $\frac{1}{8}\alpha$. Then from the four closed intervals remaining, each of length $\frac{1}{4}(1 - \frac{1}{2}\alpha - \frac{1}{4}\alpha)$, remove the middle open intervals each of length $\alpha/32$. From the eight closed intervals remaining, each of length $\frac{1}{8}(1 - \frac{1}{2}\alpha - \frac{1}{4}\alpha - \frac{1}{8}\alpha)$, remove middle open intervals each of length $\alpha/128$. After n stages the measure of the union of the open intervals removed is $\alpha(\frac{1}{2} + \frac{1}{4} + \cdots + 2^{-n})$, and therefore the measure of the union of the open intervals removed in the entire sequence of removal operations is α. The measure of the remaining Cantor set is $1 - \alpha$. For this reason, Cantor sets defined in this fashion are often called **Cantor sets of positive measure.** They are all perfect nowhere dense sets. It will be shown in Example 23, below, that all Cantor sets, of positive or zero measure, are homeomorphic (cf. Introduction, Chapter 12). It will follow, then, from the second paragraph of Example 2, above, that *every Cantor set has cardinality $\mathfrak{c}$ equal to that of $\mathfrak{R}$.*

A third construction of a Cantor set is the following: Let $0 < \beta < 1$ and let $\{\beta_n\}$ be a sequence of positive numbers such that $\sum_{n=0}^{+\infty} 2^n \beta_n = \beta$. Delete from $[0, 1]$ the open interval I_0, centered at $\frac{1}{2}$ and of length β_0. Then from $[0, 1] \setminus I_0$, delete two open intervals I_1^1, I_1^2, each centered in one of the two disjoint closed intervals whose union is $[0, 1] \setminus I_0$ and each of length β_1. Continue deletions as in the preceding constructions: At the nth stage of deletion, 2^n open intervals, I_n^1, I_n^2, $\cdots$, $I_n^{2^n}$, properly centered in the closed intervals constituting the residue at the $(n - 1)$st stage and each of length β_n, are deleted, $n = 1, 2, \cdots$.

5. A perfect nowhere dense set of irrational numbers.

A final example of a perfect nowhere dense set can be constructed by making use of a sequence $\{r_n\}$ whose terms constitute the set of all rational numbers of $(0, 1)$. Start as in the definition of the Cantor set C, but extend the open interval so that the center remains at $\frac{1}{2}$, so that its endpoints are irrational, and so that the enlarged open interval contains the point r_1. At the second stage remove from each of the two remaining closed intervals an enlarged open middle "third" in such a way that the midpoints remain midpoints, the endpoints are irrational, and the second rational number r_2 is removed. If this process is repeated according to the indicated pattern, a perfect nowhere dense set D remains, and since all rational numbers between 0 and 1 have been removed, this "Cantor" set D consists entirely of irrational numbers, except for the two points 0 and 1. If the endpoints of the original interval are chosen to be irrational numbers, a perfect nowhere dense set can be constructed in this fashion so that it consists entirely of irrational numbers.

6. A dense open set whose complement is not of measure zero.

Let A be a Cantor set of positive measure in $[0, 1]$, and let $B \equiv A' = \mathfrak{R} \setminus A$. Then B is a dense open set whose complement A has positive measure.

7. A set of the second category.

A set is said to be of the **first category** iff it is a countable union of nowhere dense sets. Any subset of a set of the first category is a set of the first category, and any countable union of sets of the first category is a set of the first category. The set $\mathbb{Q}$ of rational numbers is of the first category. A set is said to be of the **second category** iff it is not of the first category. An example of a set of the second category is the set $\mathfrak{R}$ of all real numbers. More generally, any complete metric space is of the second category (cf. [36], p. 338, Ex. 33). This general result is due to R. Baire (cf. [1], p. 108, [20], pp. 138–145, and [27], p. 204). It follows from this that the set $\mathfrak{R} \setminus \mathbb{Q}$ of irrational numbers is of the second category. We outline now a proof — independent of the general theorem just cited — that any set A of real numbers with nonempty interior $I(A)$ is of the second category. Assume the contrary, and let C be a nonempty closed interval $[a, b]$, interior to A, and let $C = F_1 \cup F_2 \cup \cdots$, where the sets F_n are *closed* and nowhere dense, $n = 1, 2, \cdots$. Let C_1 be a closed interval $[a_1, b_1] \subset (a, b) \setminus F_1$; let C_2 be a closed interval $[a_2, b_2] \subset (a_1, b_1) \setminus F_2$; in general, for $n > 1$, let $C_n = [a_n, b_n] \subset (a_{n-1}, b_{n-1}) \setminus F_n$. Then there exists a point p belonging to *every* C_n, $n = 1, 2, \cdots$ (cf. [34], p. 201, Ex. 30), whence $p \in C$. But this is impossible since p belongs to *no* F_n, $n = 1, 2, \cdots$. (Contradiction.)

8. A set that is not an F_σ set.

Recall (Example 23, Chapter 2) that an F_σ set is a set that is a countable union of closed sets. Examples of F_σ sets abound: finite sets, closed intervals, open intervals (for example, $(0, 1)$ is the union of the sets $[1/n, (n - 1)/n]$), half-open intervals, $\mathbb{Q}$ (if the rational numbers are arranged in a sequence $r_1, r_2, \cdots$, then $\mathbb{Q}$ is the union of the one-point closed sets $\{r_1\}, \{r_2\}, \cdots, \{r_n\}, \cdots$). An example of a set that is *not* an F_σ set is the set $\mathcal{R} \setminus \mathbb{Q}$ of irrational numbers. To prove this, assume the contrary, and let $\mathcal{R} \setminus \mathbb{Q} = C_1 \cup C_2 \cup \cdots$, where C_n is closed, $n = 1, 2, \cdots$. Since no subset of the set $\mathcal{R} \setminus \mathbb{Q}$ of irrational numbers has an interior point, every closed subset of $\mathcal{R} \setminus \mathbb{Q}$ is nowhere dense, and this implies that $\mathcal{R} \setminus \mathbb{Q}$ is of the first category. (Contradiction; cf. Example 7.)

9. A set that is not a G_δ set.

A set A is said to be a G_δ set iff it is a countable intersection of open sets. It follows from the **de Morgan laws** for set-complementation:

$$\left(\bigcup_{n=1}^{+\infty} A_n \right)' = \bigcap_{n=1}^{+\infty} A_n', \qquad \left(\bigcap_{n=1}^{+\infty} A_n \right)' = \bigcup_{n=1}^{+\infty} A_n'$$

that a set A is a G_δ set iff its complementary set $A' = \mathcal{R} \setminus A$ is an F_σ set. Therefore, since the set $\mathcal{R} \setminus \mathbb{Q}$ of irrational numbers is not an F_σ set, the set $\mathbb{Q}$ of rational numbers is not a G_δ set.

If countable unions of G_δ sets and countable intersections of F_σ sets are formed, two new classes of sets are obtained, called $G_{\delta\sigma}$ sets and $F_{\sigma\delta}$ sets, respectively. In fact, two infinite sequences of such classes exist: labeled $F_\sigma, F_{\sigma\delta}, F_{\sigma\delta\sigma}, \cdots$ and $G_\delta, G_{\delta\sigma}, G_{\delta\sigma\delta}, \cdots$. For a treatment of these sets, cf. [20].

10. A set A for which there exists no function having A as its set of points of discontinuity.

Let A be the set $\mathcal{R} \setminus \mathbb{Q}$ of irrational numbers. Then since A is not an F_σ set there is no real-valued function of a real variable whose set of points of discontinuity is A (cf. the final remark of Example 23, Chapter 2). In other words, there is no function from $\mathcal{R}$ to $\mathcal{R}$ that is continuous at every rational point and discontinuous at every irrational point. (Cf. Example 15, Chapter 2.)

11. A nonmeasurable set.

The axiom of choice provides a means of constructing a set that is not Lebesgue-measurable. In fact, the set thus produced cannot be measurable with respect to *any* nontrivial countably additive translation-invariant measure. More specifically, if μ is a measure function defined for all sets A of real numbers, finite-valued for bounded sets, and such that

$$\mu(x + A) = \mu(A)$$

for every $x \in \mathcal{R}$ and $A \subset \mathcal{R}$, then $\mu(A) = 0$ for every $A \subset \mathcal{R}$. We shall now prove this fact.

We start with an equivalence relation $\sim$ defined on $(0, 1] \times (0, 1]$ as follows: $x \sim y$ iff $x - y \in \mathbb{Q}$. By means of $\sim$ the half-open interval $(0, 1]$ is partitioned into disjoint equivalence classes C. The axiom of choice, applied to this family of equivalence classes, produces a set A having the two properties: (1) no two distinct points of A belong to the same equivalence class C; (2) every equivalence class C contains a point of A. In terms of the equivalence relation $\sim$ these two properties take the form: (1) no two distinct members of A are equivalent to each other; (2) every point x of $(0, 1]$ is equivalent to some member of A. We now define, for each $r \in (0, 1]$, an operation on the set A, called **translation modulo 1,** as follows:

$$(r + A) \,(\mathrm{mod}\ 1) \equiv [(r + A) \cup ((r - 1) + A)] \cap (0, 1]$$

$$= \{(r + A) \cap (0, 1]\} \cup \{((r - 1) + A) \cap (0, 1]\}.$$

The two properties of the set A stated above imply, for translation modulo 1: (1) any two sets $(r + A)(\mathrm{mod}\ 1)$ and $(s + A)(\mathrm{mod}\ 1)$ for distinct rational numbers r and s of $(0, 1]$ are disjoint; (2) every real number x of $(0, 1]$ is a member of a set $(r + A)(\mathrm{mod}\ 1)$ for some rational number r of $(0, 1]$. In other words, the half-open interval $(0, 1]$ is the union of the pair-wise disjoint countable collection $\{(r + A)(\mathrm{mod}\ 1)\}$, where $r \in \mathbb{Q} \cap (0, 1]$. An important property of the sets obtained from A by translation modulo 1 (on the basis of the assumptions made concerning μ) is that they all have the same measure as A:

$\mu((r + A)(\bmod 1))$

$$= \mu((r + A) \cap (0, 1]) + \mu(((r - 1) + A) \cap (0, 1])$$

$$= \mu((r + A) \cap (0, 1]) + \mu((r + A) \cap (1, 2])$$

$$= \mu((r + A) \cap (0, 2]) = \mu(r + A) = \mu(A).$$

On the assumption that A has positive measure we infer from the countable additivity of μ:

$$\mu((0, 1]) = \sum_{r \in Q \cap (0,1]} \mu((r + A)(\bmod 1)) = \sum_{r \in Q \cap (0,1]} \mu(A) = +\infty,$$

which is impossible since $(0, 1]$ is bounded. Consequently $\mu(A) = 0$, and

$$\mu((0, 1]) = \sum_{r \in Q \cap (0,1]} \mu((r + A)(\bmod 1)) = \sum_{r \in Q \cap (0,1]} \mu(A) = 0,$$

whence

$$\mu(\Re) = \sum_{n=-\infty}^{+\infty} \mu((n, n + 1]) = \sum_{n=-\infty}^{+\infty} \mu((0, 1]) = 0.$$

As a consequence of this, μ is the trivial measure function for which every set has measure zero.

Finally, since Lebesgue measure is a nontrivial translation-invariant measure for which bounded intervals have positive finite measure, the detailed steps just presented show that *the set A is not Lebesgue-measurable.*

Since all F_σ sets and all G_δ sets are Borel sets, and therefore measurable, the preceding nonmeasurable set is an example of a set that is neither an F_σ set nor a G_δ set.

The construction just described can be looked at in terms of sets on a circle, as follows: In the complex plane $\mathfrak{C}$, let the unit circle $\mathfrak{I} \equiv \{z \mid z \in \mathfrak{C}, \mid z \mid = 1\}$ be regarded as a group under multiplication. For each $z \in \mathfrak{I}$ there is a unique $\theta, 0 \leq \theta < 1$, such that $z = e^{2\pi i \theta}$. Let $\mathfrak{I}_0 \equiv \{z \mid z = e^{2\pi i \theta}, \theta \in \mathfrak{Q}, 0 \leq \theta < 1\}$. Then $\mathfrak{I}_0$ is a normal subgroup, and the quotient group $\mathfrak{S} = \mathfrak{I}/\mathfrak{I}_0$ exists. If $\check{\mathfrak{S}}$ is a one-to-one preimage of $\mathfrak{S}$ in $\mathfrak{I}$ (a complete set of representatives in $\mathfrak{I}$), obtained by the use of the axiom of choice, and if Lebesgue measure μ on $[0, 1)$

is carried over to a measure $\bar{\mu}$ on $\mathfrak{I}$ by the rule:

$$E \subset \mathfrak{I} \text{ is } \textbf{measurable} \text{ iff } F \equiv \{\theta \mid e^{2\pi i\theta} \in E, 0 \leqq \theta < 1\}$$

$$\text{is Lebesgue-measurable and } \bar{\mu}(E) \equiv 2\pi\mu(F),$$

then $\mathfrak{S}$ is not measurable. Indeed, $\bigcup_{\theta \in \mathfrak{Q} \cap [0,1)} e^{2\pi i\theta}\mathfrak{S}$ is a countable disjoint union of sets, each of which is measurable, all with the same measure, if $\mathfrak{S}$ is. Furthermore, this union is $\mathfrak{I}$ since $\mathfrak{S}$ is a complete set of representatives, whence $\mathfrak{I}$ is a countable disjoint union of measurable sets, all of the same measure, if $\mathfrak{S}$ if measurable. Since $\bar{\mu}(\mathfrak{I}) = 2\pi$, we see $\bar{\mu}(\mathfrak{S})$ cannot be positive. But if $\bar{\mu}(\mathfrak{S}) = 0$, then $\bar{\mu}(\mathfrak{I}) = 0$.

The procedures outlined above can be extended to more general topological groups, for example to compact groups having countably infinite normal subgroups. (For definitions and discussions of group, normal subgroup, etc., cf. [22]. Similarly, for topological groups cf. [29].)

12. A set D such that for every measurable set A, $\mu_*(D \cap A) = 0$ and $\mu^*(D \cap A) = \mu(A)$.

A set D having this property may be thought of, informally, as being the ultimate in nonmeasurability — D is as nonmeasurable as a set can be! The set D, as is the case with the nonmeasurable set A of Example 11, is again constructed with the aid of the axiom of choice, but the details are somewhat more complicated. For a complete discussion, see [18], p. 70. This example shows that every measurable set A contains a subset whose inner measure is equal to zero and whose outer measure is equal to the measure of A. It also shows that *every set of positive measure contains a nonmeasurable subset*.

F. Galvin has given a construction of a family $\{E_t\}$, $0 \leqq t \leqq 1$, of pairwise disjoint subsets of [0, 1] such that each has outer measure 1.

13. A set A of measure zero such that every real number is a point of condensation of A.

A point p is a **point of condensation** of a set A if and only if every neighborhood of p contains an uncountable set of points of A. Let C be the Cantor set of Example 1, and for any closed interval $[\alpha, \beta]$, where $\alpha < \beta$, define the set $C(\alpha, \beta)$:

$$C(\alpha, \beta) \equiv \{\alpha + (\beta - \alpha)x \mid x \in C\}.$$

Then $C(\alpha, \beta)$ is a perfect nowhere dense set of measure zero. Let B be the set that is the union of all sets $C(\alpha, \beta)$ for rational α and β, where $\alpha < \beta$. Since B is the union of a countable family of sets of measure zero it is also a set of measure zero. On the other hand, since *every* open interval I must contain a set $C(\alpha, \beta)$, and since *every* $C(\alpha, \beta)$ is uncountable, every real number must be a point of condensation of B. (Cf. [14], p. 287.)

14. A nowhere dense set A of real numbers and a continuous mapping of A onto the closed unit interval [0, 1].

The set A can be *any* Cantor set (Examples 1 and 4), since all Cantor sets are homeomorphic (Example 23). We shall describe a specific mapping ϕ for *the* Cantor set of Example 1. The mapping is that described in the second paragraph of Example 2: For any $x \in C$, let $0.c_1 c_2 c_3 \cdots$ be its ternary expansion, where $c_n = 0$ or 2, $n = 1, 2,$ $\cdots$, and let

$$\phi(x) \equiv 0 \cdot \frac{c_1}{2}\, \frac{c_2}{2}\, \frac{c_3}{2} \, \cdots \, ,$$

where the expansion on the right is now interpreted as a *binary* expansion in terms of the digits 0 and 1. It is clear that the image of C, under ϕ, is a subset of [0, 1]. To see that $[0, 1] \subset \phi(C)$, we choose an arbitrary $y \in [0, 1]$ and a binary expansion of y:

$$y = 0.b_1 b_2 b_3 \cdots .$$

Then

$$x \equiv 0.(2b_1)(2b_2)(2b_3) \cdots$$

(evaluated in the ternary system) is a point of C such that $\phi(x) = y$. Continuity of ϕ is not difficult to establish, but it is more conveniently seen in geometric terms as discussed in the following example, where the mapping ϕ is extended to a continuous mapping on the entire unit interval [0, 1].

It should be noted that ϕ is not one-to-one. Indeed, this would be impossible since C and [0, 1] are not homeomorphic, and any one-to-one continuous mapping of one compact set onto another is a homeomorphism (cf. [34], p. 192). (The set C is totally disconnected, having only one-point subsets that are connected, whereas its entire image

95

[0, 1] is connected.) An example of two points of C that have the same image is the pair $0.022000000\cdots$ and $0.020222222\cdots$, since their images are 0.011000000 and $0.010111111\cdots = 0.011000000$. In fact, two points x_1 and x_2 of C have the same image, under ϕ , *if and only if* they have the form:

$$x_1 = 0.c_1c_2\cdots c_n2000\cdots \text{ and } x_2 = 0.c_1c_2\cdots c_n0222\cdots .$$

In other words, $\phi(x_1) = \phi(x_2)$ iff x_1 and x_2 are endpoints of one of the open intervals deleted from [0, 1] in the construction of C. Therefore ϕ is an increasing function on C, and *strictly* increasing except for such pairs of endpoints. (Cf. Example 30.)

The following general theorem is an extension of the preceding "existence theorem," indicating what is possible in metric spaces for both continuous and homeomorphic (topological) images of the Cantor set (actually, of *any* Cantor set, by Example 23) and its subsets (for definitions, cf. the Introduction to Chapter 12): *Every separable metric space is a continuous image of a subset of the Cantor set. Every compact metric space is a continuous image of the Cantor set. Every compact totally disconnected metric space is a homeomorphic image of a closed subset of the Cantor set. Every compact totally disconnected perfect metric space is a homeomorphic image of the Cantor set.* (Cf. [1], pp. 119–122.)

15. A continuous monotonic function with a vanishing derivative almost everywhere.

The function defined in Example 14 can be extended so that its domain is the entire unit interval [0, 1], as follows: If $x \in [0, 1] \setminus C$, then x is a member of one of the open intervals (a, b) removed from [0, 1] in the construction of C, and therefore $\phi(a) = \phi(b)$; define $\phi(x) \equiv \phi(a) = \phi(b)$. In other words ϕ is defined to be constant on the closure of each interval removed in forming C. On the interval $[\frac{1}{3}, \frac{2}{3}]$, $\phi(x) = \frac{1}{2}$. On the intervals $[\frac{1}{9}, \frac{2}{9}]$ and $[\frac{7}{9}, \frac{8}{9}]$, the values of ϕ are $\frac{1}{4}$ and $\frac{3}{4}$, respectively. On the intervals $[\frac{1}{27}, \frac{2}{27}]$, $[\frac{7}{27}, \frac{8}{27}]$, $[\frac{19}{27}, \frac{20}{27}]$, and $[\frac{25}{27}, \frac{26}{27}]$, the values of ϕ are $\frac{1}{8}$, $\frac{3}{8}$, $\frac{5}{8}$, and $\frac{7}{8}$, respectively. If this process is repeated indefinitely, we see that the function ϕ with domain [0, 1] is increasing there, and (locally) constant in *some* neighborhood of *every* point of [0, 1] $\setminus C$. (Cf. Fig. 4.) Since ϕ is increasing on [0, 1] and since the range of ϕ is the entire interval

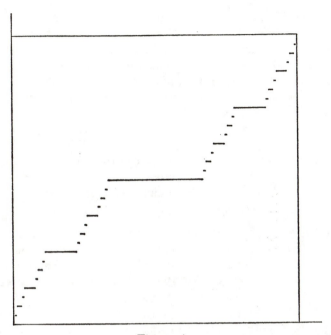

Figure 4

[0, 1], ϕ has no jump discontinuities. Since a monotonic function can have no discontinuities other than jump discontinuities (cf. [34], p. 52, Ex. 24), ϕ is continuous. Since ϕ is locally constant on the *open* subset $[0, 1] \setminus C$, which has measure 1, $\phi'(x) = 0$ almost everywhere in $[0, 1]$. The function ϕ just defined is called **the Cantor function**.

In much the same way as *the* Cantor function is defined in terms of *the* Cantor set, other "Cantor functions" can be defined in terms of other Cantor sets (of positive measure). Perhaps the simplest way, in terms of a given Cantor set A on $[0, 1]$, to define the corresponding Cantor function g is to define g first on the closures of the successively removed intervals: on the central interval, $g(x) \equiv \frac{1}{2}$; on the next two the values of g are $\frac{1}{4}$ and $\frac{3}{4}$, on the next four the values are $\frac{1}{8}, \frac{3}{8}, \frac{5}{8}, \frac{7}{8}$, etc. On the dense subset $[0, 1] \setminus A$, g is increasing and its range is dense in $[0, 1]$. Therefore the domain of g can be extended to $[0, 1]$ so that g is increasing and continuous on $[0, 1]$, with range $[0, 1]$.

By making use of Example 5 — a "Cantor set" of irrational

numbers — it is possible to construct a function h that is increasing and continuous on [0, 1], with range [0, 1], and such that $h'(x) = 0$ for *every* rational number $x \in$ [0, 1]. In fact, the range of $h(x)$ for rational $x \in$ [0, 1] can be made equal to the entire set $Q \cap$ [0, 1] of rational points of [0, 1], instead of simply those of the form $m/2^n$ as in the preceding cases. In this way we obtain a function satisfying the requirements of Example 11g, Chapter 1.

For a *strictly* monotonic example, see Example 30.

16. A topological mapping of a closed interval that destroys both measurability and measure zero.

If ϕ is the Cantor function of Example 15, define a function ψ by:

$$\psi(x) \equiv x + \phi(x), \qquad 0 \leqq x \leqq 1,$$

with range [0, 2]. Since ϕ is increasing on [0, 1], and continuous there, ψ is *strictly* increasing and topological there (continuous and one-to-one with a continuous inverse on the range of ψ). Since each open interval removed from [0, 1] in the construction of the Cantor set C is mapped by ψ onto an interval of [0, 2] of equal length, $\mu(\psi(I \setminus C)) = \mu(I \setminus C) = 1$, whence $\mu(\psi(C)) = 1$. Since C is a set of measure zero, ψ is an example of a topological mapping that maps a set of measure zero onto a set of positive measure. Now let D be a nonmeasurable subset of $\psi(C)$ (cf. Example 12). Then $\psi^{-1}(D)$ is a subset of the set C of measure zero, and is therefore also a (measurable) set of measure zero. Thus ψ is an example of a topological mapping that maps a measurable set onto a nonmeasurable set. (See also Example 23.)

17. A measurable non-Borel set.

The set $\psi^{-1}(D)$ of Example 16 is a measurable set, but since it is the image under a topological mapping of a non-Borel set D, $\psi^{-1}(D)$ is not a Borel set. (Cf. [20], p. 195.)

18. Two continuous functions that do not differ by a constant but that have everywhere identical derivatives (in the finite or infinite sense).

This example was given by Rey Pastor [38] (also cf. [10], p. 133).*

---

* The example given by Pastor is in error. For a correct example see S. Saks, *Theory of the integral*, Warsaw (1937), pp. 205–206.

Let ϕ be the Cantor function of Example 15. On the unit interval $[0, 1]$ define the function $h(x)$ to be equal to zero on the Cantor set C, and on each open interval (a, b) removed in the construction of C define $h(x)$ so that its graph consists of two congruent semicircles with diameter on the x axis, one semicircle lying above the x axis on the left-hand half of (a, b) and the other semicircle lying below the x axis on the right-hand half of (a, b):

$$h(x) \equiv \begin{cases} \left[\left(\dfrac{b-a}{4}\right)^2 - \left(x - \dfrac{3a+b}{4}\right)^2\right]^{1/2} & \text{if } a < x \leq \dfrac{a+b}{2}, \\ -\left[\left(\dfrac{b-a}{4}\right)^2 - \left(x - \dfrac{a+3b}{4}\right)^2\right]^{1/2} & \text{if } \dfrac{a+b}{2} \leq x < b. \end{cases}$$

Then h is everywhere continuous on the interval $[0, 1]$. Finally, let $f(x) \equiv 2\phi(x) + h(x)$ and $g(x) \equiv \phi(x) + h(x)$. Then $f'(x) = g'(x)$ for $0 \leq x \leq 1$: for every x of the Cantor set C, $f'(x) = g'(x) = +\infty$; for every x that is the midpoint of an interval removed in the formation of C, $f'(x) = g'(x) = -\infty$; for every other $x \in [0, 1] \setminus C$, $f'(x) = g'(x) = h'(x)$. On the other hand, $f(x) - g(x) = \phi(x)$, and $\phi(x)$ is not a constant function.

19. A set in $[0, 1]$ of measure 1 and category I.

First example: Let A_n be a Cantor set in $[0, 1]$ of measure $(n-1)/n$, $n = 1, 2, \cdots$, and let $A \equiv A_1 \cup A_2 \cup \cdots$. Then, since A_n is nowhere dense, for $n = 1, 2, \cdots$, A is of the first category. On the other hand, since

$$\mu(A_n) = \frac{n-1}{n} \leq \mu(A) \leq 1$$

for $n = 1, 2, \cdots$, $\mu(A) = 1$.

Second example: A second such set is given by the complement (relative to the unit interval) of the set of the second example under Example 20, below.

20. A set in $[0, 1]$ of measure zero and category II.

First example: If A is the set of the first example under Example 19, above, then its complement $A' = [0, 1] \setminus A$ is of the second category (if it were of the first category, the interval $[0, 1]$ would be a union of two sets of the first category and hence also of the first category), and of measure zero ($\mu([0, 1] \setminus A) + \mu(A) = 1$).

Second example: Let $\mathbb{Q} \cap [0, 1]$ be the range of a sequence $\{r_n\}$, and for each pair of positive integers k and n, let I_{kn} be an open interval containing r_n and of length $< 2^{-k-n}$. If $A_k \equiv \bigcup_{n=1}^{+\infty} I_{kn}$, and $B_k \equiv [0, 1] \setminus A_k$, then A_k is an open set containing $\mathbb{Q} \cap [0, 1]$ and having measure $\mu(A_k) < 2^{-k}$, and hence B_k is a compact nowhere dense set of measure $\mu(B_k) > 1 - 2^{-k}$. (The measure of A_k is less than or equal to the sum of the lengths of the intervals I_{kn}, $n = 1, 2, \cdots$, and B_k can have no interior points since it consists of irrational points only.) Therefore the set $B \equiv \bigcup_{k=1}^{+\infty} B_k$ is a subset of $[0, 1]$ of measure 1 and of the first category; the set $A \equiv \bigcap_{k=1}^{+\infty} A_k = [0, 1] \setminus B$ is a subset of $[0, 1]$ of measure zero and of the second category.

21. A set of measure zero that is not an F_σ set.

First example: The first example under Example 20 cannot be a countable union of closed sets $F_1, F_2, \cdots$, since if it were, each F_n would be a closed set of measure zero and therefore nowhere dense. But this would mean that the set under consideration would be of the first category. (Contradiction.)

Second example: The second example under Example 20, for the same reasons as those just given, has the stated properties.

Third example: The non-Borel set of Example 17 is of measure zero, but cannot be an F_σ set since every F_σ set is a Borel set.

22. A set of measure zero such that there is no function — Riemann-integrable or not — having the set as its set of points of discontinuity.

Each set under Example 21 is such a set, since for any function on $\mathfrak{R}$ into $\mathfrak{R}$ the set of points of discontinuity is an F_σ set (cf. Example 23, Chapter 2; also cf. Example 8, Chapter 4).

The present example is of interest in connection with the theorem: *A necessary and sufficient condition for a real-valued function defined and bounded on a compact interval to be Riemann-integrable there, is that the set of its points of discontinuity be of measure zero* (cf. [36], p. 153, Ex. 54). A careless reading of this theorem might lead one to believe that *every* set of measure zero is involved, since the condition of the theorem is *both* necessary *and* sufficient.

23. Two perfect nowhere dense sets in [0, 1] that are homeomorphic, but exactly one of which has measure zero.

We shall prove even more: If C is the Cantor set on [0, 1] of measure zero and if A is *any* Cantor set on [0, 1] of positive measure, then there exists a homeomorphism f of [0, 1] onto [0, 1] such that $f(C) = A$. An immediate consequence of this will be the corollary that *all Cantor sets are homeomorphic.*

The idea of the mapping is similar to that of the original Cantor function (Example 15): Arrange the intervals I_1, I_2, $\cdots$ and the intervals J_1, J_2, $\cdots$ deleted from [0, 1] in the formation of C and A, respectively, in the same "order sense." That is, let I_1 and J_1 be the middle intervals first removed; then let I_2 and J_2 be the "left middles" and I_3 and J_3 be the "right middles" next removed, etc. Then map the closure of I_n onto the closure of J_n linearly and increasingly, for $n = 1, 2, \cdots$. Then f is defined and strictly increasing on a dense subset of [0, 1], and since its range is also dense on [0, 1], the continuous extension of f to [0, 1] is immediate, as described in the second paragraph under Example 15. (Cf. Fig. 5.) The present function f is a second example of the type called for in Example 16.

24. Two disjoint nonempty nowhere dense sets of real numbers such that every point of each set is a limit point of the other.

Let A be any Cantor set in [0, 1], and let B be the subset of A consisting of all endpoints of the open intervals that were deleted from [0, 1] in the construction of A, and let $E \equiv A \setminus B$. Then B and E satisfy the requirements.

If the containing space is not restricted to being $\Re$, examples are easily constructed. For instance, in the Euclidean plane two sets satisfying the stated conditions are the set of rational numbers on the x axis and the set of irrational numbers on the x axis.

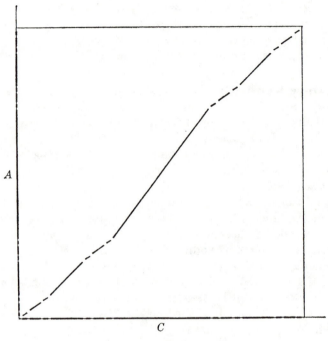

Figure 5

25. Two homeomorphic sets of real numbers that are of different category.

We start by defining an increasing continuous function on [0, 1] that is similar to a Cantor function as defined in the second paragraph under Example 15. In the present instance let $\{J_n\}$ be the sequence of open sets removed from [0, 1] in the construction of a Cantor set A, as described in Example 23, and let $\{s_n\}$ be a one-to-one mapping of $\mathfrak{N}$ onto $\mathbb{Q} \cap (0, 1)$. Define the sequence $\{r_n\}$ as follows: let $r_1 \equiv s_1$; let $r_2 \equiv s_n$ where n is the least positive integer such that $s_n < r_1$; let $r_3 \equiv s_n$ where n is the least positive integer such that $s_n > r_1$. Then let $r_4 \equiv s_n$ where n is the least positive integer such that $s_n < r_2$; let $r_5 \equiv s_n$ where n is the least positive integer such that $r_2 < s_n < r_1$; let $r_6 \equiv s_n$ where n is the least positive integer such that $r_1 < s_n < r_3$; let $r_7 \equiv s_n$ where n is the least positive integer such that $s_n > r_3$. If this procedure is continued, the rational numbers between 0 and 1

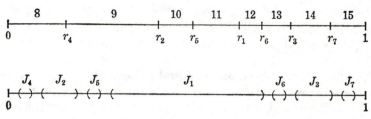

Figure 6

are arranged in a sequence $\{r_n\}$ in such a way that their order relation corresponds to that of the sequence of intervals J_n, as indicated in Figure 6. In other words, $r_m < r_n$ if and only if J_m lies to the left of J_n. We now define the function f so that $f(x) \equiv r_n$ if x belongs to the closure $\bar{J}_n$ of J_n, $n = 1, 2, \cdots$. As in Example 15, f is defined and increasing on a dense subset of $[0, 1]$, with range dense in $[0, 1]$, and can be extended to a continuous increasing function mapping $[0, 1]$ onto $[0, 1]$. If B and E are defined as in Example 24, then f maps B onto the set $Q \cap (0, 1)$, and E onto the set $(0, 1) \setminus Q$ of all irrational numbers between 0 and 1. In this latter case, the mapping between E and $(0, 1) \setminus Q$ is strictly increasing and *bicontinuous*. (The continuity of the inverse mapping follows from the order-relationship among the points of E and the correspondence to that order-relationship among the points of $(0, 1) \setminus Q$.) Therefore E and $(0, 1) \setminus Q$ are homeomorphic. That is, *any Cantor set shorn of its "endpoints" is homeomorphic to the set of irrational numbers between 0 and 1.* But E is nowhere dense and hence of the first category, whereas $(0, 1) \setminus Q$ is of the second category (cf. Example 7).

It should be noted that, in contrast to the sets of Example 23, the homeomorphism of the present Example cannot be induced from a homeomorphism between containing intervals. (If two spaces are homeomorphic and if two sets correspond under this homeomorphism, then if one is nowhere dense so is the other; if one is of the first category, then so is the other.)

26. Two homeomorphic sets of real numbers such that one is dense and the other is nowhere dense.

If, in Example 25, the rational numbers $\{r_n\}$ are not restricted to the interval $(0, 1)$ but are permitted to encompass the entire set Q, then a homeomorphism between E and the set $Q' = \Re \setminus Q$ of *all* irrational numbers is obtained. The set E is nowhere dense, and the set Q' is dense. (Cf. the final paragraph under Example 25.)

27. A function defined on $\Re$, equal to zero almost everywhere and whose range on every nonempty open interval is $\Re$.

We shall arrive at the construction of a function f having the stated properties in stages. Our first goal is to define a function g on the open interval $(0, 1)$ that maps the set $C \cap (0, 1)$, where C is the Cantor set of measure zero, onto $\Re$. If ϕ is the Cantor function (Example 15), then g can be defined:

$$g(x) \equiv \tan\left[\pi(\phi(x) - \tfrac{1}{2})\right], \qquad 0 < x < 1.$$

The second step is to define, for an arbitrary open interval $I = (a, b)$, a subset Z_I of measure zero and a function g_I with domain Z_I and range $\Re$. This can be done as follows:

$$Z_I \equiv \{a + (b - a)x \mid x \in C \cap (0, 1)\},$$

$$g_I(x) \equiv g\left(\frac{x - a}{b - a}\right), \qquad x \in Z_I.$$

We start defining the desired function f by letting it be equal to zero on the set $\mathcal{I}$ of integers. We now define a sequence $\{U_n\}$ of open sets as follows: Let $U_1 \equiv \Re \setminus \mathcal{I}$, which is the union of all open intervals of the form $(n, n + 1)$, where n is an integer. In *each* of these intervals I let Z_I be the set of measure zero defined above, and on the set Z_I define f to be equal to g_I. The subset U_2 of U_1 where f has not yet been defined is an open set, and therefore a union of disjoint open intervals. In *each* of these intervals I let Z_I be the set of measure zero defined above, and on the set Z_I define f to be equal to g_I. The subset U_3 of U_2 where f has not yet been defined is again an open set, and therefore a union of disjoint open intervals. If the sets Z_I are again defined as above, the domain of the function f can once more be extended, to include these sets of measure zero. Let this process continue, by means of a sequence $\{U_n\}$ of open sets each of which has a complement of measure zero. The function f thereby becomes defined on a set of measure zero — the complement of the intersection of $U_1, U_2, \cdots$, or equivalently, the union of their complements $U_1', U_2', \cdots$ — in such a way that *every* nonempty open interval must contain one of the open intervals I of one of the open sets U_n, and therefore a set Z_I on which the range of f is $\Re$. Finally, we define f to be identically zero where it has not already been defined.

28. A function on $\Re$ whose graph is dense in the plane.
The function of Example 27 has this property.

29. A function f such that $0 \leqq f(x) < +\infty$ everywhere but $\int_a^b f(x)\, dx = +\infty$ for every nonempty open interval (a, b).
A function having these properties can be constructed by repeating the procedures of Example 27, with the following two changes: (i) let the set C be replaced by a Cantor set of measure $\frac{1}{2}$ on $[0, 1]$ and (ii) define the function g_I:

$$g_I(x) \equiv \frac{1}{|I|^2}\, \chi(Z_I),$$

where $|I|$ denotes the length of the interval I and $\chi(A)$ is the characteristic function of the set A (cf. the Introduction, Chapter 1). Each set Z_I has measure $\frac{1}{2}|I|$, and therefore the integral of g_I over I is equal to $1/(2|I|)$. Since *every* nonempty interval of the form (c, d) contains subintervals I of arbitrarily small length, the integral $\int_c^d f(x)\, dx$ is arbitrarily large and hence, being a constant, is equal to $+\infty$.

30. A continuous strictly monotonic function with a vanishing derivative almost everywhere. (See page 195 for elaboration.)
A function f with these properties is given by A. C. Zaanen and W. A. J. Luxemburg [3], as follows: If ϕ is the Cantor function of Example 15, let $\psi(x) \equiv \phi(x)$ if $x \in [0, 1]$ and $\psi(x) \equiv 0$ if $x \in \Re \setminus [0, 1]$, let $\{[a_n, b_n]\}$ be the sequence of closed intervals $[0, 1]$, $[0, \frac{1}{2}]$, $[\frac{1}{2}, 1]$, $[0, \frac{1}{4}]$, $[\frac{1}{4}, \frac{1}{2}]$, $\cdots$, $[0, \frac{1}{8}]$, $\cdots$, and define

$$f(x) \equiv \sum_{n=1}^{+\infty} 2^{-n} \psi\left(\frac{x - a_n}{b_n - a_n}\right) \quad \text{for} \quad 0 \leqq x \leqq 1.$$

31. A bounded semicontinuous function that is not Riemann-integrable, nor equivalent to a Riemann-integrable function.
The characteristic function f of a Cantor set A of positive measure on $[0, 1]$ is bounded and everywhere upper semicontinuous. Since its

set of points of discontinuity is A, which has positive measure, f is not Riemann-integrable on $[0, 1]$. Two functions are **equivalent** iff they are equal almost everywhere. If the values of f are changed on a set of measure zero, the resulting function also has a set of positive measure as its set of points of discontinuity.

32. A bounded measurable function not equivalent to a Riemann-integrable function.
The function of Example 31 has this property.

33. A bounded monotonic limit of continuous functions that is not Riemann-integrable, nor equivalent to a Riemann-integrable function. (Cf. Example 10, Chapter 4.)
The function f of Example 31 can be obtained as the limit of a decreasing sequence $\{f_n\}$ of continuous functions as follows: For any open interval $I = (a, b)$, where $0 \leqq a < b \leqq 1$, and for any positive integer n, define the function $g_{n,I}$:

$$g_{n,I}(x) \equiv \begin{cases} 1 & \text{if} \quad 0 \leqq x \leqq a, \\[2mm] 1 & \text{if} \quad b \leqq x \leqq 1, \\[2mm] 0 & \text{if} \quad a + \dfrac{b-a}{2^n} \leqq x \leqq b - \dfrac{b-a}{2^n}, \\[2mm] \text{linear if} & a \leqq x \leqq a + \dfrac{b-a}{2^n}, \\[2mm] \text{linear if} & b - \dfrac{b-a}{2^n} \leqq x \leqq b. \end{cases}$$

If $\{J_n\}$ is the sequence of open intervals removed from $[0, 1]$ in the formation of the Cantor set A of positive measure (cf. Example 23), define the sequence $\{f_n\}$:

$$f_1 \equiv g_{1,J_1},$$
$$f_2 \equiv g_{2,J_1} \cdot g_{2,J_2},$$
$$\cdots$$
$$f_n \equiv g_{n,J_1} \cdot g_{n,J_2} \cdot \cdots \cdot g_{n,J_n}.$$

34. A Riemann-integrable function f, and a continuous function g, both defined on $[0, 1]$ and such that the composite

function $f(g(x))$ is not **Riemann-integrable on [0, 1], nor equivalent to a Riemann-integrable function there.** (Cf. Example 9, Chapter 4.)

The function of Example 31 is of the required form $f(g(x))$ if $f(x)$ is defined to be 0 for $0 \le x < 1$ and 1 for $x = 1$, and if $g(x)$ is defined to be 1 if $x \in A$, and $1 - \frac{1}{2}(b - a) + | x - \frac{1}{2}(a + b) |$ if x belongs to an interval $I = (a, b)$ removed from [0, 1] in the formation of A. The function $g(x)$ is continuous since for all x_1 and x_2 of [0, 1], $| g(x_1) - g(x_2) | \le | x_1 - x_2 |$.

Note that by appropriate use of "bridging functions" (cf. Example 12, Chapter 3) the functions $g_{n,I}$ of Example 33 and the portions defining the function g of Example 34 — and therefore the functions f_n of Example 33 and the complete function g of Example 34 — may be replaced by infinitely differentiable functions.

Finally, it should be noted that in the reverse order this example is impossible. In other words, every continuous function (with a compact interval as domain) of a Riemann-integrable function is Riemann-integrable. (Cf. [36], p. 153, Ex. 55.)

35. A bounded function possessing a primitive on a closed interval but failing to be Riemann-integrable there.

Let the function g be defined for positive x (cf. Example 2, Chapter 3) by the formula $g(x) \equiv x^2 \sin(1/x)$, and for any positive number c, let x_c be the greatest positive x less than or equal to c such that $g'(x) = 0$. For any positive c define the function g_c for $0 < x \le c$:

$$g_c(x) \equiv \begin{cases} g(x) & \text{if } 0 < x \le x_c, \\ g(x_c) & \text{if } x_c \le x \le c. \end{cases}$$

If A is any Cantor set of positive measure on [0, 1], define the function f as follows: If $x \in A$ let $f(x) \equiv 0$, and if x belongs to an interval $I = (a, b)$ removed from [0, 1] in the formation of A, let

$$f(x) \equiv \begin{cases} g_c(x - a) & \text{if } a < x \le \frac{1}{2}(a + b), \\ g_c(-x + b) & \text{if } \frac{1}{2}(a + b) \le x < b, \end{cases}$$

where $c \equiv \frac{1}{2}(b - a)$.

If x is any point of A and if y is any *other* point of [0, 1], then either $f(y) = 0$ or y is a point of some removed interval $I = (a, b)$. In the

107

former case, $|(f(y) - f(x))/(y - x)| = 0 < |y - x|$. In the latter case, let d be defined to be the endpoint of (a, b) nearer y. Then, with $c \equiv \frac{1}{2}(b - a)$,

$$\left|\frac{f(y) - f(x)}{y - x}\right| = \left|\frac{f(y)}{y - x}\right| \leqq \left|\frac{f(y)}{y - d}\right|$$

$$= \left|\frac{g_c(|y - d|)}{y - d}\right| \leqq \left|\frac{|y - d|^2}{y - d}\right| = |y - d| \leqq |y - x|.$$

Therefore, in either case, $|(f(y) - f(x))/(y - x)| \leqq |y - x|$, and consequently $f'(x) = 0$ for every $x \in A$.

On the other hand, if x belongs to any removed interval (a, b),

$$|f'(x)| \leqq |2z \sin (1/z) - \cos (1/z)| \leqq 3,$$

for some z between 0 and 1, so that f is everywhere differentiable on $[a, b]$, and its derivative f' is bounded there.

Finally since $\overline{\lim}_{y \to 0+} g'(y) = 1$ (cf. Introduction, Chapter 2), it follows that for any point x of A, $\overline{\lim}_{y \to x} f'(y) = 1$. Therefore the function f' is discontinuous at every point of A, and hence on a set of of positive measure. The function f' therefore satisfies all conditions specified in the statement above.

A construction similar to the above was (presumably first) given by the Italian mathematician V. Volterra (1860–1940) in *Giorn. di Battaglini* 19(1881), pp. 353–372.

36. A function whose improper (Riemann) integral exists but whose Lebesgue integral does not.

If $f(x) \equiv \sin x/x$ if $x \neq 0$ and $f(0) \equiv 1$, then the improper integral $\int_0^{+\infty} f(x)\, dx$ converges conditionally (cf. [34] p. 465). That is, the integral converges, but $\int_0^{+\infty} |f(x)|\, dx = +\infty$. This means that the function $|f(x)|$ is not Lebesgue-integrable on $[0, +\infty)$, and therefore neither is $f(x)$.

37. A function that is Lebesgue-measurable but not Borel-measurable.

The characteristic function of the Lebesgue-measurable non-Borel set in Example 17 satisfies these conditions.

38. A measurable function $f(x)$ and a continuous function $g(x)$ such that the composite function $f(g(x))$ is not measurable.

In the notation of Example 16, let $E \equiv \psi^{-1}(D)$. Then the characteristic function $f \equiv \chi_E$ of the set E is measurable and $g \equiv \psi^{-1}$ is continuous, but the composite function $f(g(x))$ is the nonmeasurable characteristic function of the nonmeasurable set D.

It should be noted that in the reverse order this example is impossible. In other words, every continuous function of a measurable function is measurable.

39. A continuous monotonic function $g(x)$ and a continuous function $f(x)$ such that

$$\int_0^1 f(x) \, dg(x) \neq \int_0^1 f(x) g'(x) \, dx.$$

Let $f(x) \equiv 1$ on $[0, 1]$, and let g be the Cantor function ϕ of Example 15. Then the Riemann-Stieltjes (cf. [36], p. 179) or Lebesgue-Stieltjes integral (cf. [18], [30], and [32]) on the left above is equal to $\phi(1) - \phi(0) = 1$, while the Lebesgue integral on the right is equal to 0 since the integrand is almost everywhere equal to 0.

40. Sequences of functions converging in different senses.

If $f, f_1, f_2, \cdots$ are Lebesgue-integrable functions on either the unit interval $[0, 1]$ (more generally, on a measurable set of finite measure) or the real number system $\mathfrak{R}$ (more generally, on a measurable set of infinite measure), then there are many senses in which the statement

$$\lim_{n \to +\infty} f_n = f$$

may be interpreted. We shall consider here four specific meanings, indicate the implications among them, and give counterexamples when such implications are absent. We shall indicate by the single letter S either $[0, 1]$ or $\mathfrak{R}$ as the domain for the functions concerned. The four interpretations for the limit statement given above that we shall consider are:

I. Functions of a Real Variable

(*i*) **Convergence almost everywhere*:**

$$\lim_{n \to +\infty} f_n(x) = f(x) \text{ for almost every } x \in S.$$

(*ii*) **Convergence in measure:**

$$\forall \, \varepsilon > 0 \, , \, \lim_{n \to +\infty} \mu\{x \, | \, | f_n(x) - f(x) \, | > \varepsilon\} = 0.$$

(*iii*) **Mean convergence†:**

$$\lim_{n \to +\infty} \int_S | f_n(x) - f(x) | \, dx = 0.$$

(*iv*) **Dominated convergence‡:**
Convergence almost everywhere holds, and there exists a Lebesgue-integrable function g such that $| f_n(x) | \leq | g(x) |$ for $n = 1, 2, \cdots$ and almost all $x \in S$.

We start with two statements concerning the implications that hold among (*i*)–(*iv*). If S is of finite measure, then

$$(iv) \Rightarrow \left\{ \begin{array}{c} (i) \\ (iii) \end{array} \right\} \Rightarrow (ii).$$

If S is of infinite measure, then

$$(iv) \Rightarrow \left\{ \begin{array}{l} (i) \\ (iii) \Rightarrow (ii). \end{array} \right.$$

(Cf. [18], [30], and [32].)

Examples now follow to show that all of the implications missing above may fail. For all but the last one, each of the examples serves for spaces of either finite or infinite measure, since all of the functions involved are zero for $x \in \Re \setminus [0, 1]$.

---

* A closely related type of convergence is *convergence everywhere*, which has a rather trivial relationship to (*i*).

† This is the same as convergence in the Banach space L^1 of all integrable functions, reduced modulo functions almost everywhere equal to zero. This can be generalized by means of an exponent $p \geq 1$ to convergence in the Banach space L^p of measurable functions the pth powers of whose absolute values are integrable, reduced modulo functions almost everywhere equal to zero. For further discussion, see [16], [18], [29], and [32].

‡ The type of dominated convergence needed, in case (*iii*) is replaced by convergence in L^p, is that given by $| f_n(x) | \leq | g(x) |$, where $g \in L^p$.

$(i) \nRightarrow (iii)$: Let $f(x) \equiv 0$ for all $x \in \Re$,

$$f_n(x) \equiv \begin{cases} n & \text{if } 0 < x < 1/n, \\ 0 & \text{if } x \in \Re \setminus (0, 1/n), \end{cases}$$

for $n = 1, 2, \cdots$.

$(i) \nRightarrow (iv)$: Same as $(i) \nRightarrow (iii)$ since $(iv) \Rightarrow (iii)$.

$(iii) \nRightarrow (i)$: Let $f(x) \equiv 0$ for all $x \in \Re$. For each $n \in \mathfrak{N}$, write $n = 2^k + m$, where $0 \leqq m < 2^k$, $k = 0, 1, 2, \cdots$; then k and m are uniquely determined by n. Let

$$f_n(x) \equiv \begin{cases} 1 & \text{if } \dfrac{m}{2^k} \leqq x \leqq \dfrac{m+1}{2^k}, \\ \\ 0 & \text{otherwise for } x \in \Re. \end{cases}$$

Then $\int_S |f_n(x) - f(x)| \, dx = 2^{-k} \to 0$ as $n \to +\infty$, but $\lim_{n \to +\infty} f_n(x)$ exists for *no* $x \in [0, 1]$.

$(iii) \nRightarrow (iv)$: Same as $(iii) \nRightarrow (i)$ since $(iv) \Rightarrow (i)$.

$(ii) \nRightarrow (i)$: Same as $(iii) \nRightarrow (i)$, since for the function f_n of that example, and any positive ε,

$$\mu\{x \mid |f_n(x) - f(x)| > \varepsilon\} \leqq 2^{-k} \to 0 \quad \text{as} \quad n \to +\infty.$$

$(ii) \nRightarrow (iii)$: Let $f(x) \equiv 0$ for all $x \in \Re$, and for any $n \in \mathfrak{N}$ let k and m be determined as in the example for $(iii) \nRightarrow (i)$. Let

$$f_n(x) \equiv \begin{cases} 2^k & \text{if } \dfrac{m}{2^k} \leqq x \leqq \dfrac{m+1}{2^k}, \\ \\ 0 & \text{otherwise for } x \in \Re. \end{cases}$$

Then for any positive number ε,

$$\mu\{x \mid |f_n(x) - f(x)| > \varepsilon\} \leqq 2^{-k} \to 0 \quad \text{as} \quad n \to +\infty,$$

but $\displaystyle \int_S |f_n(x) - f(x)| \, dx = 1 \nrightarrow 0 \quad \text{as} \quad n \to +\infty$.

$(ii) \nRightarrow (iv)$: Same as $(ii) \nRightarrow (i)$ or $(ii) \nRightarrow (iii)$, since $(iv) \Rightarrow (i)$ and $(iv) \Rightarrow (iii)$.

Finally, we give an example where the space S is the space $\Re$ of infinite measure:

$(i) \nRightarrow (ii)$: Let $f(x) \equiv 0$ for all $x \in \Re$,

$$f_n(x) \equiv \begin{cases} 1 & \text{if } n \leqq x \leqq n + 1, \\ \\ 0 & \text{otherwise for } x \in \Re. \end{cases}$$

41. Two measures μ and ν on a measure space (X, S) such that μ is absolutely continuous with respect to ν and for which no function f exists such that $\mu(E) = \int_E f(x)\, d\nu(x)$ for all $E \in S$.

Let $X \equiv \mathcal{R}$ and let S be the class of all subsets of X. For any set $E \in S$ define

$$\mu(E) \equiv \begin{cases} 0 & \text{if } E \text{ is countable,} \\ +\infty & \text{if } E \text{ is uncountable,} \end{cases}$$

$$\nu(E) \equiv \begin{cases} n & \text{if } E \text{ consists of } n \text{ points, } n \geq 0, \\ +\infty & \text{if } E \text{ is infinite.} \end{cases}$$

Then $\nu(E) = 0 \Rightarrow E = \emptyset$ and hence $\mu(E) = 0$, so μ is absolutely continuous with respect to ν. On the other hand, if f is a function such that

$$\mu(E) = \int_E f(x)\, d\nu(x)$$

for all sets E, then this equation holds in particular when E is an arbitrary one-point set, $E = \{y\}$, in which case

$$\mu(E) = 0 = \int_E f(x)\, d\nu(x) = f(y).$$

But this means that the function f is identically 0, and consequently that $\mu(E) = 0$ for every $E \in S$. (Contradiction.)

If we interpret $\pm \infty \cdot 0$ as 0, the following statement is true: *If f is a nonnegative extended-real-valued function measurable with respect to a measure function ν on a measure space (X, S) and if*

$$\mu(E) \equiv \int_E f(x)\, d\nu(x)$$

for all measurable sets E, then μ is a measure function on (X, S) that is absolutely continuous with respect to ν. The preceding counterexample shows that the unrestricted converse of this statement is false. The Radon-Nikodym theorem (cf. [18]) is a restricted form of the converse.

Part II
Higher Dimensions

Chapter 9
Functions of Two Variables

Introduction

In this chapter a basic familiarity with the concepts of continuity and differentiability of functions of two variables — and for the last two examples, line integrals, simple-connectedness, and vector analysis — will be assumed. If $f(x, y)$ is a differentiable function of the two variables x and y, its partial derivatives will be alternatively denoted:

$$\frac{\partial f}{\partial x} = f_x(x, y) = f_1(x, y), \qquad \frac{\partial f}{\partial y} = f_y(x, y) = f_2(x, y),$$

$$\frac{\partial^2 f}{\partial x^2} = f_{xx}(x, y) = f_{11}(x, y), \qquad \frac{\partial^2 f}{\partial x \partial y} = f_{xy}(x, y) = f_{12}(x, y), \cdots.$$

A **region** is a nonempty open set R any two of whose points can be connected by a broken line segment lying completely in R.

1. A discontinuous function of two variables that is continuous in each variable separately.

Let the function $f(x, y)$ with domain $\mathfrak{R} \times \mathfrak{R}$ be defined:

$$f(x, y) = \begin{cases} \dfrac{xy}{x^2 + y^2} & \text{if } x^2 + y^2 \neq 0, \\ 0 & \text{if } x = y = 0. \end{cases}$$

Then f is discontinuous at the origin since arbitrarily near $(0, 0)$ there exist points of the form (a, a) at which f has the value $\frac{1}{2}$. On the other hand, for any fixed value of y (whether zero or nonzero), the function $g(x) \equiv f(x, y)$ is everywhere a continuous function of x. For a similar reason $f(x, y)$ is a continuous function of y for every fixed value of x.

2. A function of two variables possessing no limit at the origin but for which any straight line approach gives the limit zero.

Let the function $f(x, y)$ with domain $\Re \times \Re$ be defined:

$$f(x, y) \equiv \begin{cases} \dfrac{x^2 y}{x^4 + y^2} & \text{if} \quad x^2 + y^2 \neq 0, \\ 0 & \text{if} \quad x = y = 0, \end{cases}$$

and let L be an arbitrary straight line through the origin. If L is either coordinate axis, then on $L, f(x, y)$ is identically zero, and hence has the limit 0 as $(x, y) \to (0, 0)$ along L. If L is the line $y = mx$, then on $L, f(x, y)$ has the value

$$f(x, mx) = \frac{mx^3}{x^4 + m^2 x^2} = \frac{mx}{x^2 + m^2}$$

for $x \neq 0$. Therefore $\lim_{x \to 0} f(x, mx) = 0$. In spite of this fact, $f(x, y)$ is discontinuous at $(0, 0)$ since, arbitrarily near $(0, 0)$, there are points of the form (a, a^2) at which f has the value $\frac{1}{2}$.

3. A refinement of the preceding example.

Let the function $f(x, y)$ with domain $\Re \times \Re$ be defined:

$$f(x, y) \equiv \begin{cases} \dfrac{e^{-1/x^2} y}{e^{-2/x^2} + y^2} & \text{if} \quad x \neq 0, \\ 0 & \text{if} \quad x = 0, \end{cases}$$

and let C be an arbitrary curve through the origin and of the form $x^m = (y/c)^n$ or $y = cx^{m/n}$, where m and n are relatively prime positive integers and c is a nonzero constant ($x \geq 0$ in case n is even). Then if the point (x, y) is permitted to approach $(0, 0)$ along C, we have:

$$\lim_{x \to 0} f(x, cx^{m/n}) = \lim_{x \to 0} \frac{ce^{-1/x^2} x^{-m/n}}{e^{-2/x^2} x^{-2m/n} + c^2} = 0.$$

(Cf. Example 10, Chapter 3.) In spite of the fact that the limit of $f(x, y)$ as (x, y) approaches the origin along an arbitrary algebraic curve of the type $y = cx^{m/n}$ is zero, the function $f(x, y)$ is discontinuous at $(0, 0)$ since there are points of the form $(a, e^{-1/a^2})$ arbitrarily near $(0, 0)$ at which f has the value $\frac{1}{2}$.

4. A discontinuous (and hence nondifferentiable) function of two variables possessing first partial derivatives everywhere.
Each function of the three preceding examples has these properties.

5. Functions f for which exactly two of the following exist and are equal:

$$\lim_{(x,y) \to (a,b)} f(x, y), \qquad \lim_{x \to a} \lim_{y \to b} f(x, y), \qquad \lim_{y \to b} \lim_{x \to a} f(x, y).$$

Let the three limits written above be designated (i), (ii), and (iii), respectively. The following functions are such that the indicated limit does not exist but the other two do and are equal:

(i): Example 1, with $(a, b) = (0, 0)$.

(ii): $f(x, y) \equiv \begin{cases} y + x \sin (1/y) & \text{if } y \neq 0, \\ 0 & \text{if } y = 0, \end{cases}$

with $(a, b) = (0, 0)$.

(iii): $f(x, y) \equiv \begin{cases} x + y \sin (1/x) & \text{if } x \neq 0, \\ 0 & \text{if } x = 0, \end{cases}$

with $(a, b) = (0, 0)$.

In both examples (ii) and (iii),

$$|f(x, y)| \leq |x| + |y| \leq 2(x^2 + y^2)^{1/2},$$

and hence $\lim_{(x,y) \to (0,0)} f(x, y) = 0$. Each iterated limit, (ii) or (iii) that exists is equal to 0.

It should be noted that if both limits (i) and (ii) exist they must be equal, and that if both limits (i) and (iii) exist they must be equal (cf. [34], p. 184).

6. Functions f for which exactly one of the following exists:

$$\lim_{(x,y) \to (a,b)} f(x, y), \qquad \lim_{x \to a} \lim_{y \to b} f(x, y), \qquad \lim_{y \to b} \lim_{x \to a} f(x, y).$$

As in Example 5, designate the three limits above by (i), (ii), and (iii), respectively. The following functions are such that the indicated limit exists but the other two do not:

(i): $f(x, y) \equiv \begin{cases} x \sin(1/y) + y \sin(1/x) & \text{if } xy \neq 0, \\ 0 & \text{if } xy = 0, \end{cases}$

with $(a, b) = (0, 0)$.

(ii): $f(x, y) \equiv \begin{cases} \dfrac{xy}{x^2 + y^2} + y \sin \dfrac{1}{x} & \text{if } x \neq 0, \\ 0 & \text{if } x = 0, \end{cases}$

with $(a, b) = (0, 0)$.

(iii): $f(x, y) \equiv \begin{cases} \dfrac{xy}{x^2 + y^2} + x \sin \dfrac{1}{y} & \text{if } y \neq 0, \\ 0 & \text{if } y = 0, \end{cases}$

with $(a, b) = (0, 0)$.

7. A function f for which $\lim_{x \to a} \lim_{y \to b} f(x, y)$ and $\lim_{y \to b} \lim_{x \to a} f(x, y)$ exist and are unequal.

If $\qquad f(x, y) \equiv \begin{cases} \dfrac{x^2 - y^2}{x^2 + y^2} & \text{if } x^2 + y^2 \neq 0, \\ 0 & \text{if } x = y = 0, \end{cases}$

then

$$\lim_{x \to 0} \lim_{y \to 0} f(x, y) = \lim_{x \to 0} (x^2/x^2) = 1,$$

$$\lim_{y \to 0} \lim_{x \to 0} f(x, y) = \lim_{y \to 0} (-y^2/y^2) = -1.$$

8. A function $f(x, y)$ for which $\lim_{y \to 0} f(x, y) = g(x)$ exists uniformly in x, $\lim_{x \to 0} f(x, y) = h(y)$ exists uniformly in y, $\lim_{x \to 0} g(x) = \lim_{y \to 0} h(y)$, but $\lim_{(x,y) \to (0,0)} f(x, y)$ does not exist.

Let $\qquad f(x, y) \equiv \begin{cases} 1 & \text{if } xy \neq 0, \\ 0 & \text{if } xy = 0. \end{cases}$

Then

$$g(x) \equiv \lim_{y \to 0} f(x, y) = \begin{cases} 1 & \text{if } x \neq 0, \\ 0 & \text{if } x = 0. \end{cases}$$

$$h(y) \equiv \lim_{x \to 0} f(x, y) = \begin{cases} 1 & \text{if } y \neq 0, \\ 0 & \text{if } y = 0, \end{cases}$$

and both of these limits are uniform over the entire real number system. However, since there are points arbitrarily near $(0, 0)$ at which f is equal to 0, and points arbitrarily near $(0, 0)$ at which f is equal to 1, the limit of $f(x, y)$ as $(x, y) \to (0, 0)$ cannot exist.

It should be noted that by the Moore-Osgood theorem (cf. [36], p. 313), the present counterexample is impossible if all points of the form $(0, y)$ and all points of the form $(x, 0)$ are excluded from the domain of f.

9. A differentiable function of two variables that is not continuously differentiable.

If

$$f(x, y) \equiv \begin{cases} x^2 \sin(1/x) + y^2 \sin(1/y) & \text{if } xy \neq 0, \\ x^2 \sin(1/x) & \text{if } x \neq 0 \text{ and } y = 0, \\ y^2 \sin(1/y) & \text{if } x = 0 \text{ and } y \neq 0, \\ 0 & \text{if } x = y = 0, \end{cases}$$

then both functions

$$f_x(x, y) = \begin{cases} 2x \sin(1/x) - \cos(1/x) & \text{if } x \neq 0, \\ 0 & \text{if } x = 0, \end{cases}$$

$$f_y(x, y) = \begin{cases} 2y \sin(1/y) - \cos(1/y) & \text{if } y \neq 0, \\ 0 & \text{if } y = 0 \end{cases}$$

are discontinuous at the origin and hence f is not continuously differentiable there. However, f is differentiable everywhere. For example, f is differentiable at $(0, 0)$, since for $h^2 + k^2 \neq 0$, $f(h, k) - f(0, 0)$ can be written in the form

$$f_x(0, 0)h + f_y(0, 0)k + \varepsilon_1(h, k)h + \varepsilon_2(h, k)k,$$

where

$$\lim_{(h,k) \to (0,0)} \varepsilon_1(h, k) = \lim_{(h,k) \to (0,0)} \varepsilon_2(h, k) = 0.$$

Indeed, this representation takes the specific form

$$f(h, k) - f(0, 0) = \begin{cases} \left(h \sin \dfrac{1}{h}\right) h + \left(k \sin \dfrac{1}{k}\right) k & \text{if} \quad hk \neq 0, \\[2ex] \left(h \sin \dfrac{1}{h}\right) h + 0 \cdot k & \text{if} \quad h \neq 0 \quad \text{and} \quad k = 0, \\[2ex] 0 \cdot h + \left(k \sin \dfrac{1}{k}\right) k & \text{if} \quad h = 0 \quad \text{and} \quad k \neq 0. \end{cases}$$

10. A differentiable function with unequal mixed second order partial derivatives.

If

$$f(x, y) \equiv \begin{cases} xy \dfrac{x^2 - y^2}{x^2 + y^2} & \text{if} \quad x^2 + y^2 \neq 0, \\[2ex] 0 & \text{if} \quad x = y = 0, \end{cases}$$

then

$$f_y(x, 0) = \begin{cases} x & \text{if} \quad x \neq 0, \\[2ex] \lim_{k \to 0} \dfrac{f(0, k)}{k} = 0 & \text{if} \quad x = 0, \end{cases}$$

$$f_x(0, y) = \begin{cases} -y & \text{if} \quad y \neq 0, \\[2ex] \lim_{h \to 0} \dfrac{f(h, 0)}{h} = 0 & \text{if} \quad y = 0, \end{cases}$$

and hence at the origin,

$$f_{xy}(0, 0) = \lim_{h \to 0} \frac{f_y(h, 0) - f_y(0, 0)}{h} = \lim_{h \to 0} \frac{h}{h} = 1,$$

$$f_{yx}(0, 0) = \lim_{k \to 0} \frac{f_x(0, k) - f_x(0, 0)}{k} = \lim_{k \to 0} \frac{-k}{k} = -1.$$

The function f is continuously differentiable since both $\partial f/\partial x$ and $\partial f/\partial y$ are continuous everywhere. In particular, $\partial f/\partial x$ is continuous at the origin since, for $x^2 + y^2 \neq 0$,

$$\left| \frac{\partial f}{\partial x} \right| = \frac{|x^4 y + 4x^2 y^3 - y^5|}{(x^2 + y^2)^2} \leq \frac{6(x^2 + y^2)^{5/2}}{(x^2 + y^2)^2} = 6(x^2 + y^2)^{1/2}.$$

The present example would be impossible in the presence of *continuity* of the mixed partial derivatives f_{xy} and f_{yx} in a neighborhood of the origin. In fact (cf. [34], p. 2 63) *if f, f_x, and f_y exist in a region R and if f_{xy} (or f_{yx}) exists and is continuous at any point (a, b) of R, then f_{yx} (or f_{xy}) also exists at (a, b) and $f_{xy} = f_{yx}$ there.*

11. A continuously differentiable function f of two variables x and y, and a plane region R such that $\partial f/\partial y$ vanishes identically in R but f is not independent of y in R.

Let L be the ray (closed half-line) in $\Re \times \Re$:

$$L \equiv \{(x, y) \mid x \geqq 0, y = 0\},$$

and let R be the region $(\Re \times \Re) \setminus L$. The function

$$f(x, y) \equiv \begin{cases} x^3 & \text{if } x > 0 \text{ and } y > 0, \\ 0 & \text{otherwise for } (x, y) \in R, \end{cases}$$

is continuously differentiable in R and, in fact, has continuous second order partial derivatives. (If x^3 is replaced by e^{-1/x^2}, f has continuous partial derivatives of all orders.) Although the first partial derivative $f_2(x, y)$ of f with respect to y vanishes identically throughout R, the function f is *not* independent of y; for instance, $f(1, 1) = 1$ and $f(1, -1) = 0$. This example demonstrates the invalidity of the following argument in showing that a function f having identically vanishing first partial derivatives throughout a region R is constant there (cf. [34], p. 280): "Since $\partial f/\partial x = 0$, f does not depend on x; since $\partial f/\partial y = 0$, f does not depend on y; therefore f depends on neither x nor y and must be a constant." If the intersection with a region R of every line parallel to the y axis is an interval, the present counter example becomes impossible (cf. [34], p. 288, Ex. 32).

12. A locally homogeneous continuously differentiable function of two variables that is not homogeneous.

A function $f(x, y)$ is **homogeneous** of degree n in a region R iff for all x, y, and positive λ such that both (x, y) and $(\lambda x, \lambda y)$ are in R, $f(\lambda x, \lambda y) = \lambda^n f(x, y)$. A function $f(x, y)$ is **locally homogeneous** of degree n in a region R iff f is homogeneous of degree n in some neighborhood of every point of R.

Let L be the ray (closed half-line) in $\Re \times \Re$:

$$L \equiv \{(x, y) \mid x = 2, y \geqq 0\},$$

and let R be the region $(\mathfrak{R} \times \mathfrak{R}) \setminus L$. The function

$$f(x, y) \equiv \begin{cases} y^4/x & \text{if } x > 2 \text{ and } y > 0, \\ y^3 & \text{otherwise for } (x, y) \in R, \end{cases}$$

is continuously differentiable in R (in fact, f has continuous second order partial derivatives). Since, for λ near 1 and for any $(x, y) \in R$, $f(\lambda x, \lambda y) = \lambda^3 f(x, y)$, f is locally homogeneous of degree 3 in R. However, f is not homogeneous of degree 3 in R since, for the point $(x, y) = (1, 2)$ and for $\lambda = 4$, $f(x, y) = 8$ and $f(4x, 4y) = f(4, 8) = 1024 \neq 4^3 \cdot 8$. The function f is not homogeneous of degree n for any $n \neq 3$ since if it were it would be locally homogeneous of degree n, which is clearly impossible.

13. A differentiable function of two variables possessing no extremum at the origin but for which the restriction to an arbitrary line through the origin has a strict relative minimum there.

The function

$$f(x, y) \equiv (y - x^2)(y - 3x^2)$$

has no relative extremum at the origin since there are points of the form $(0, b)$ arbitrarily near the origin at which f is positive, and also points of the form $(a, 2a^2)$ arbitrarily near the origin at which f is negative. If the domain of f is restricted to the x axis, the restricted function $3x^4$ has a strict absolute minimum at $x = 0$. If the domain of f is restricted to the y axis, the restricted function y^2 has a strict absolute minimum at $y = 0$. If the domain of f is restricted to the line $y = mx$ through the origin where $0 < |m| < +\infty$, the restricted function of the parameter x:

$$g(x) = f(x, mx) = (mx - x^2)(mx - 3x^2) = m^2x^2 - 4mx^3 + 3x^4$$

has a strict relative minimum at the origin since $g'(0) = 0$ and $g''(0) = 2m^2 > 0$.

14. A refinement of the preceding example.

The function

$$f(x, y) \equiv \begin{cases} (y - e^{-1/x^2})(y - 3e^{-1/x^2}) & \text{if } x \neq 0, \\ y^2 & \text{if } x = 0 \end{cases}$$

has no relative extremum at the origin (cf. Example 13), but if the domain of f is restricted to the algebraic curve $y = cx^{m/n}$, where m and n are relatively prime positive integers and c is a nonzero constant ($x \geqq 0$ in case n is even), the restricted function of the parameter x:

$$g(x) = f(x, cx^{m/n}) = (cx^{m/n} - e^{-1/x^2})(cx^{m/n} - 3e^{-1/x^2})$$

$$= x^{2m/n}[c^2 - 4ce^{-1/x^2}x^{-m/n} + 3e^{-2/x^2}x^{-2m/n}]$$

has a strict relative minimum at $x = 0$. This is true since the factor $x^{2m/n}$ is positive for $x \neq 0$, while the quantity in brackets has the positive limit c^2 as $x \to 0$.

15. A function f for which $d/dx \int_a^b f(x, y)\, dy \neq \int_a^b [\partial/\partial x f(x, y)]\, dy$, although each integral is proper.

The function

$$f(x, y) \equiv \begin{cases} \dfrac{x^3}{y^2} e^{-x^2/y} & \text{if } y > 0, \\ 0 & \text{if } y = 0, \end{cases}$$

with domain the closed upper half-plane $y \geqq 0$, is a continuous function of x for each fixed value of y and a continuous function of y for each fixed value of x, although as a function of the two variables x and y it is discontinuous at $(0, 0)$ (let $y = x^2$). By explicit integration,

$$g(x) \equiv \int_0^1 f(x, y)\, dy = xe^{-x^2}$$

for every real number x (including $x = 0$), and hence $g'(x) = e^{-x^2}(1 - 2x^2)$ for every real number x (including $x = 0$). For $x \neq 0$,

$$\int_0^1 f_1(x, y)\, dy = \int_0^1 e^{-x^2/y}\left(\frac{3x^2}{y^2} - \frac{2x^4}{y^3}\right) dy = e^{-x^2}(1 - 2x^2),$$

while for $x = 0$, since $f_1(0, y) = 0$ for all y (including $y = 0$):

$$\int_0^1 f_1(0, y)\, dy = \int_0^1 0\, dy = 0.$$

Therefore,

$$g'(0) = 1 \neq \int_0^1 f_1(0, y) \, dy = 0.$$

Each integral evaluated above is proper since in every case the integrand is a continuous function of the variable of integration.

16. A function *f* for which $\int_0^1 \int_0^1 f(x, y) \, dy \, dx \neq \int_0^1 \int_0^1 f(x, y) \, dx \, dy$, although each integral is proper.

Let

$$f(x, y) \equiv \begin{cases} y^{-2} & \text{if} \quad 0 < x < y < 1, \\ -x^{-2} & \text{if} \quad 0 < y < x < 1, \\ 0 & \text{otherwise if} \quad 0 \leq x \leq 1, \quad 0 \leq y \leq 1. \end{cases}$$

For $0 < y < 1$,

$$\int_0^1 f(x, y) \, dx = \int_0^y \frac{dx}{y^2} - \int_y^1 \frac{dx}{x^2} = 1,$$

and therefore

$$\int_0^1 \int_0^1 f(x, y) \, dx \, dy = \int_0^1 1 \, dy = 1.$$

Similarly, for $0 < x < 1$,

$$\int_0^1 f(x, y) \, dy = -\int_0^x \frac{dy}{x^2} + \int_x^1 \frac{dy}{y^2} = -1,$$

and therefore

$$\int_0^1 \int_0^1 f(x, y) \, dy \, dx = \int_0^1 (-1) \, dx = -1.$$

17. A double series $\sum_{m,n} a_{mn}$ for which $\sum_m \sum_n a_{mn} \neq \sum_n \sum_m a_{mn}$, although convergence holds throughout.

Let (a_{mn}), where m designates the number of the (horizontal) row and n designates the number of the (vertical) column, be the infinite matrix (cf. Example 20, Chapter 6):

$$\begin{pmatrix} 0 & \frac{1}{2} & \frac{1}{4} & \frac{1}{8} & \frac{1}{16} & \frac{1}{32} & \cdots \\ -\frac{1}{2} & 0 & \frac{1}{2} & \frac{1}{4} & \frac{1}{8} & \frac{1}{16} & \cdots \\ -\frac{1}{4} & -\frac{1}{2} & 0 & \frac{1}{2} & \frac{1}{4} & \frac{1}{8} & \cdots \\ -\frac{1}{8} & -\frac{1}{4} & -\frac{1}{2} & 0 & \frac{1}{2} & \frac{1}{4} & \cdots \\ \cdots\cdots\cdots\cdots\cdots\cdots\cdots\cdots\cdots \end{pmatrix}.$$

Then

$$\sum_{n=1}^{+\infty} a_{mn} = 2^{-m} + 2^{-m-1} + \cdots = 2^{-m+1}, \qquad m = 1, 2, \cdots,$$

and hence

$$\sum_{m=1}^{+\infty} \sum_{n=1}^{+\infty} a_{mn} = 1 + 2^{-1} + 2^{-2} + \cdots = 2.$$

Similarly,

$$\sum_{n=1}^{+\infty} \sum_{m=1}^{+\infty} a_{mn} = \sum_{n=1}^{+\infty} (-2^{-n+1}) = -2.$$

(Cf. [14], p. 109.)

18. A differential $P\,dx + Q\,dy$ and a plane region R in which $P\,dx + Q\,dy$ is locally exact but not exact.

The expression

$$P\,dx + Q\,dy,$$

where P and Q are continuous in a region R of $\mathfrak{R} \times \mathfrak{R}$, is called an **exact differential** in R iff there exists a differentiable function ϕ defined in R such that

$$\frac{\partial \phi}{\partial x} = P, \qquad \frac{\partial \phi}{\partial y} = Q$$

throughout R. The expression $P\,dx + Q\,dy$ is called **locally exact** in a region R iff it is exact in some neighborhood of every point of R. A necessary and sufficient condition for $P\,dx + Q\,dy$ to be exact in a region R is that for every sectionally smooth closed curve C lying in R the line integral of $P\,dx + Q\,dy$ vanishes:

$$\int_C P\,dx + Q\,dy = 0.$$

(Cf. [34], p. 587.) A necessary and sufficient condition for $P\,dx + Q\,dy$, where P and Q are continuously differentiable, to be locally exact in a region R is that at every point of R

$$\frac{\partial P}{\partial y} = \frac{\partial Q}{\partial x}.$$

The expression

$$P \, dx + Q \, dy \equiv -\frac{y}{x^2 + y^2} \, dx + \frac{x}{x^2 + y^2} \, dy$$

is locally exact throughout the "punctured plane"

$$R \equiv \{(x, y) \mid x^2 + y^2 > 0\},$$

since

$$\frac{\partial P}{\partial y} = \frac{\partial Q}{\partial x} = \frac{y^2 - x^2}{(x^2 + y^2)^2}$$

if $x^2 + y^2 > 0$. On the other hand, $P \, dx + Q \, dy$ is not exact in R, since if C is the unit circle $x = \cos \theta$, $y = \sin \theta$, $0 \leqq \theta \leqq 2\pi$, then, with θ as parameter,

$$\int_C P \, dx + Q \, dy = \int_0^{2\pi} [(-\sin \theta)(-\sin \theta) + \cos^2 \theta] \, d\theta = 2\pi \neq 0.$$

It should be noted that if R is simply-connected (cf. [34], p. 598), then $P \, dx + Q \, dy$ is exact in R iff it is locally exact in R (cf. [34], p. 601).

19. A solenoidal vector field defined in a simply-connected region and possessing no vector potential.

A vector field (cf. [34], p. 568) $P\vec{i} + Q\vec{j} + R\vec{k}$, where P, Q, and R are continuously differentiable functions over a region W in three-dimensional Euclidean space, is said to be **solenoidal** in W iff its divergence vanishes identically there:

$$\frac{\partial P}{\partial x} + \frac{\partial Q}{\partial y} + \frac{\partial R}{\partial z} = 0.$$

If a vector field $\vec{F}$ is the curl (cf. [34], p. 572) of a vector field $\vec{G}$, in a region W, the vector field $\vec{G}$ is called a **vector potential** for $\vec{F}$. Since the divergence of the curl always vanishes identically (cf. [34], p. 572), any vector field that has a vector potential is solenoidal. The converse, however, is not true, as the example

$$\vec{F} = (x^2 + y^2 + z^2)^{-3/2} (x\vec{i} + y\vec{j} + z\vec{k}),$$

for the region

$$W = \{(x, y, z) \mid x^2 + y^2 + z^2 > 0\},$$

shows. That $\vec{F}$ is solenoidal is shown by straightforward differentiation:

$$\frac{\partial}{\partial x} \{(x^2 + y^2 + z^2)^{-3/2}x\} + \frac{\partial}{\partial y} \{x^2 + y^2 + z^2)^{-3/2}y\} + \cdots$$

$$= (x^2 + y^2 + z^2)^{-5/2} [(-2x^2 + y^2 + z^2)$$

$$+ (x^2 - 2y^2 + z^2) + \cdots] = 0.$$

That $\vec{F}$ has no vector potential $\vec{G}$ can be shown by consideration of the sphere $x^2 + y^2 + z^2 = 1$. If $\vec{n}$ denotes the outer normal unit vector for this sphere S, then the surface integral $\int\int_S \vec{F} \cdot \vec{n} \, dS$ is equal to

$$\int\int_S \{(x^2 + y^2 + z^2)^{-3/2}(x\vec{i} + y\vec{j} + z\vec{k}) \cdot (x^2 + y^2 + z^2)^{-1/2}$$

$$(x\vec{i} + y\vec{j} + z\vec{k}) \, dS = \int\int_S 1 \, dS = 4\pi.$$

However, if $\vec{F}$ were the curl of a vector potential, then by Stokes's theorem (cf. [34], pp. 636, 637), the surface integral $\int\int_S \vec{F} \cdot n \, dS$ over the *closed* surface S would vanish. The region W is simply-connected (cf. [34], pp. 639, 640).

Simple-connectedness of a region can be thought of thus, that any simple closed curve in the region may be shrunk to a point without leaving the region. In the "punctured space" region W of this example, any simple closed curve not passing through the origin can be shrunk to a point without passing through the origin — and hence without leaving W. The kind of pathology for the region W that permits the present counterexample is the impossibility of shrinking spherical surfaces — or "sphere-like" surfaces — to a point without leaving W.

Chapter 10
Plane Sets

Introduction

In this chapter we shall assume that the reader is familiar with the elements of the topology of the Euclidean plane, including such ideas as boundedness, openness, closedness, compactness, denseness, and nowhere-denseness. A few other concepts are defined in the following paragraphs. In each case the space is assumed to be the Euclidean plane, E_2.

The **distance** $d(A, B)$ between two nonempty sets A and B is defined:

$$d(A, B) \equiv \inf \{d(p, q) \mid p \in A, q \in B\},$$

where $d(p, q)$ is the distance between the points $p : (x_1, y_1)$ and $q : (x_2, y_2)$, and is given by the formula $[(x_2 - x_1)^2 + (y_2 - y_1)^2]^{1/2}$. Thus, the distance between two sets is always nonnegative, is zero if the sets have a point in common, and may be zero if the sets are disjoint. If the sets are disjoint and *compact*, their distance is positive (cf. [34], p. 200 (Ex. 17)). The **diameter** $\delta(A)$ of a nonempty set A is defined

$$\delta(A) \equiv \sup \{d(p, q) \mid p \in A, q \in A\},$$

is always nonnegative, and is finite if and only if A is bounded. If A is compact its diameter is attained as the distance between two of its points (cf. [34], p. 200 (Ex. 18)).

A **closed disk** is a set of the form

$$\{(x, y) \mid (x - h)^2 + (y - k)^2 \leqq r^2\},$$

for some point (h, k) and a positive number r. An **open disk** is defined similarly, with the inclusive inequality $\leqq$ being replaced by a strict inequality $<$.

Two nonempty sets A and B are **separated** iff they are disjoint and neither contains a limit point of the other: $A \cap \bar{B} = \bar{A} \cap B = \emptyset$. A nonempty set E is **connected** iff there do not exist two nonempty separated sets A and B whose union is E. A set containing more than one point is **totally disconnected** iff its only connected subsets are one-point sets. A set A is **locally connected** iff whenever $p \in A$ and N is a neighborhood of p there exists a subneighborhood M of p such that every pair of points of M belongs to a connected subset of N.

An **arc** is a continuous mapping into E_2 of a closed interval (which may be taken to be the unit interval $[0, 1]$), or the range of such a mapping. In this latter case, when the arc is considered as a point-set, the mapping is called a **parametrization** of the arc. If the mapping is $f(t) = (x(t), y(t))$, the functions $x(t)$ and $y(t)$ are called the **parametrization functions** for the mapping. If $f(t)$, $a \leqq t \leqq b$, is an arc, and if $a = a_0 < a_1 < \cdots < a_n = b$, then the *polygonal arc* made up of the segments $f(a_0)f(a_1)$, $f(a_1)f(a_2)$, $\cdots$, $f(a_{n-1})f(a_n)$ is called an **inscribed polygon**, and the supremum of the lengths

$$d(f(a_0), f(a_1)) + d(f(a_1), f(a_2)) + \cdots + d(f(a_{n-1}), f(a_n))$$

for all inscribed polygons for the given arc is called the **length** of the arc. An arc is **rectifiable** iff its length is finite. An arc is rectifiable iff its parametrization functions are both of bounded variation (cf. [36], p. 353 (Ex. 27)). An arc $f(t)$, for $a \leqq t \leqq b$, is a **closed curve** iff $f(a) = f(b)$.

An arc $f(t)$ is **simple** iff it is a one-to-one mapping. In this case its inverse is also continuous and the mapping is a homeomorphism (cf. [34], p. 240). If $y = g(x)$ is continuous on $[a, b]$ then its graph is a simple arc (with parametrization $x = t$, $y = g(t)$, $t \in [a, b]$). A **simple closed curve** is an arc $f(t)$ such that if its domain is the closed interval $[a, b]$, then $f(t_1) = f(t_2)$ iff $t_1 = t_2$ or $\{t_1, t_2\} = \{a, b\}$. Equivalently, a simple closed curve is a homeomorphic image of a circle.

A **region** is a connected open set. The **Jordan curve theorem** states that the complement of any simple closed curve C consists of two disjoint regions for each of which C is the frontier. (Cf. [33].) A **Jordan region** is either of the two regions just described, for some simple closed curve C. A **non-Jordan region** is a region that is not a Jordan region.

If $\{C_n\}$ is a decreasing sequence of nonempty compact sets ($C_n \supset C_{n+1}$ for $n = 1, 2, \cdots$), then there exists at least one point belonging to every C_n, $n = 1, 2, \cdots$; in other words, the intersection of the C_n's is nonempty: $\bigcap_{n=1}^{+\infty} C_n \neq \emptyset$. (Cf. [34], p. 201 (Ex. 30).)

A set A is **convex** iff the closed segment joining any pair of points of A lies entirely in A. (A one-point set is considered to be a special case of a closed segment.) Since any intersection of convex sets is convex and since the plane is convex, every set in the plane is contained in a "smallest convex set," the intersection of *all* convex sets containing it. This resulting smallest convex set is called the **convex hull** of the given set. Its closure, called the **convex closure** of the set, is the smallest closed convex set containing it (cf. [36], p. 332 (Ex. 39)).

A mapping is **open** iff the image of every open set of its domain is open. A mapping is **closed** iff the image of every closed set of its domain is closed.

For some of the examples of this chapter some familiarity with plane Lebesgue measure and integration will be assumed. References to Lebesgue theory are given in the Bibliography, and cited in Chapter 8.

1. Two disjoint closed sets that are at a zero distance.

Let $F_1 \equiv \{(x, y) \mid xy = 1\}$, $F_2 \equiv \{(x, y) \mid y = 0\}$ = the x-axis. Then F_1 and F_2 are closed and disjoint. For any $\varepsilon > 0$, the points $(2/\varepsilon, \varepsilon/2)$ and $(2/\varepsilon, 0)$ in F_1 and F_2, respectively, are at a distance $\frac{1}{2}\varepsilon < \varepsilon$.

2. A bounded plane set contained in no minimum closed disk.

By a **minimum closed disk** containing a given bounded plane set A we mean a closed disk containing A and contained in every closed disk that contains A. An arbitrary two-point set is contained in no minimum disk in this sense. In contrast to this, any nonempty bounded plane set is contained in a closed disk of *minimum radius*. Any nonempty plane set A is contained in a minimum closed convex set in the sense that A is contained in a closed convex set (its convex

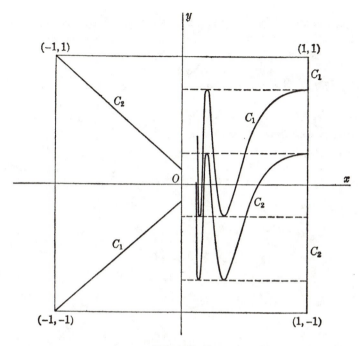

Figure 7

closure) that is itself contained in every closed convex set that contains A. In the space $\Re$ of one dimension every nonempty bounded set is contained in a minimum closed interval.

3. "Thin" connected sets that are not simple arcs.

In the present context the word "thin" means "nowhere dense in the plane."

First example: The set

$$S_1 \equiv \{(x, y) \mid y = \sin (1/x), \quad 0 < x \leqq 1\} \cup \{(0, 0)\}$$

is not a simple arc because it is not compact ($\{0\} \times [-1, 1] \subset \overline{S_1}$).

Second example: If S_1 is the set of the first example, let

$$S_2 \equiv \overline{S_1} = \{(x, y) \mid y = \sin (1/x), 0 < x \leqq 1\} \cup (\{0\} \times [-1, 1]).$$

131

Then although S_2 is compact, the removal of an arbitrary set of points from the segment $\{0\} \times [-1, 1]$ does not disconnect S_2. It will be shown in Example 11 that the set S_2 of this example is not an arc. In Example 24 we shall describe a connected set in the plane that becomes *totally* disconnected upon the removal of one point.

4. Two disjoint plane circuits contained in a square and connecting both pairs of opposite vertices.

For purposes of this example a "circuit" will mean a nowhere dense connected set. Let the square be $[-1, 1] \times [-1, 1]$, and let the circuits be given as follows (cf. Fig. 7):

$$C_1 \equiv \{(x, y) \mid y = \tfrac{7}{8}x - \tfrac{1}{8}, \quad -1 \leqq x \leqq 0\}$$

$$\cup \{(x, y) \mid y = \tfrac{1}{2} \sin (\pi/2x) + \tfrac{1}{4}, \quad 0 < x < 1\}$$

$$\cup \{(x, y) \mid x = 1, \quad \tfrac{3}{4} \leqq y \leqq 1\}.$$

$$C_2 \equiv \{(x, y) \mid y = -\tfrac{7}{8}x + \tfrac{1}{8}, \quad -1 \leqq x \leqq 0\}$$

$$\cup \{(x, y) \mid y = \tfrac{1}{2} \sin (\pi/2x) - \tfrac{1}{4}, \quad 0 < x < 1\}$$

$$\cup \{(x, y) \mid x = 1, \quad -1 \leqq y \leqq \tfrac{1}{4}\}.$$

Then C_1 connects $(-1, -1)$ to $(1, 1)$ and C_2 connects $(-1, 1)$ to $(1, -1)$, and $C_1 \cap C_2 = \emptyset$.

5. A mapping of the interval [0, 1] onto the square [0, 1] × [0, 1].

If $t \in [0, 1)$, let $0.t_1t_2t_3 \cdots$ be a binary expansion of t, and to avoid ambiguity we assume that this expansion contains infinitely many binary digits equal to 0. The point (x, y) of the unit square $S \equiv [0, 1] \times [0, 1]$ that is the image of t under the mapping f is defined

$$x \equiv 0.t_1t_3t_5 \cdots, \qquad y \equiv 0.t_2t_4t_6 \cdots.$$

Finally, define $f(1) \equiv (1, 1)$. It is not difficult to see that f is a many-to-one onto mapping. For example, the point $(0.1, 0.1)$ is the image of precisely three distinct points 0.11, $0.100101010101\cdots$, and $0.011010101\cdots$.

The mapping f is not continuous. For example, if $\{t_n\}$ is the sequence

$$0.0011, \quad 0.001111, \quad 0.00111111, \quad 0.0011111111, \quad \cdots,$$

and if $(x_n, y_n) \equiv f(t_n)$, then the sequences $\{x_n\}$ and $\{y_n\}$ are both

$$0.01, \quad 0.011, \quad 0.0111, \quad 0.01111, \quad \cdots.$$

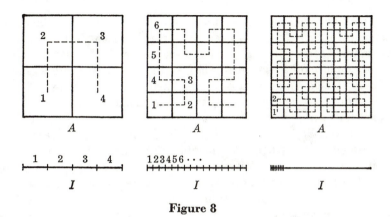

Figure 8

However, $t_n \rightarrow 0.01$, and $(x_n, y_n) \rightarrow (0.1, 0.1)$, while $f(0.01) = (0.0, 0.1) \neq (0.1, 0.1)$. That is, $\lim_{n \rightarrow +\infty} f(t_n) \neq f(\lim_{n \rightarrow +\infty} t_n)$.

It is left as an exercise for the reader to show that f is neither open (the image of the open interval from 0.101 to 0.111 contains the point $(0.1, 0.1)$ but does not contain it in its interior) nor closed (the image of the closed interval from 0.001 to 0.01 has the point $(0.1, 0.1)$ as a limit point but not a member).

6. A space-filling arc in the plane.

By a **space-filling arc** we mean an arc lying in a Euclidean space of dimension greater than one and having a nonempty interior in that space (it is *not* nowhere dense). In 1890 the Italian mathematician G. Peano (1858–1932) startled the mathematical world with the first space-filling arc. We present here a description (given in 1891 by the German mathematician D. Hilbert (1862–1943)) of an arc that fills the unit square $S = [0, 1] \times [0, 1]$. Higher-dimensional analogues can be described similarly.

As indicated in Figure 8, the idea is to subdivide S and the unit interval $I = [0, 1]$ into 4^n closed subsquares and subintervals, respectively, and to set up a correspondence between subsquares and subintervals so that inclusion relationships are preserved (at each stage of subdivision, if a square corresponds to an interval, then its subsquares correspond to subintervals of that interval).

We now define the continuous mapping f of I onto S : If $x \in I$, then at each stage of subdivision x belongs to *at least* one closed sub-

133

interval. Select either one (if there are two) and associate the corresponding square. In this way a decreasing sequence of closed squares is obtained corresponding to a decreasing sequence of closed intervals. This sequence of closed squares has the property that there is exactly one point belonging to all of them. This point is by definition $f(x)$. It remains to be shown that (i) the point $f(x)$ is well-defined, that is, independent of any choice of intervals containing x; (ii) the range of f is S; and (iii) f is continuous. The details are left to the reader.

It should be noted that the mapping f just defined is many-to-one in places. (For example, the three points $\frac{1}{6}$, $\frac{1}{2}$, and $\frac{5}{6}$ are all mapped onto the point $(\frac{1}{2}, \frac{1}{2})$.) This is inevitable, since if f were one-to-one, then it would be a homeomorphism, whereas I and S are not homeomorphic (removal of any three points disconnects I but not S). The fact that f is many-to-one is somewhat paradoxical since it seems to say that I has more points than S!

7. A space-filling arc that is almost everywhere within a countable set.

If ϕ is the Cantor function of Example 15, Chapter 8, if f is the mapping of the preceding Example 6, and if $g(x) \equiv f(\phi(x))$, then g maps the Cantor set C onto the unit square $[0, 1] \times [0, 1]$, and the complementary set $[0, 1] \setminus C$ onto the image under f of the set of points of the form $m \cdot 2^{-n}$, where n is a positive integer and m is a positive integer less than 2^n.

The present example could also be described as *a space-filling arc that is almost everywhere stationary*, or *a space-filling arc that is almost everywhere almost nowhere*.

8. A space-filling arc that is almost everywhere differentiable.

By "almost everywhere differentiable" we mean "defined by parametrization functions that are almost everywhere differentiable." The mapping defined in Example 7 has this property.

9. A continuous mapping of [0, 1] onto [0, 1] that assumes every value an uncountable number of times.

Each of the parametrization functions of the space-filling arcs of Examples 6 and 7 has this property as, indeed, must each parametrization function for any continuous mapping of $[0, 1]$ onto $[0, 1] \times [0, 1]$. Each of the parametrization functions for the mapping of Example 7 has the additional property that *it is differentiable with a vanishing derivative almost everywhere*. (Cf. [2].)

10. A simple arc in the unit square and of plane measure arbitrarily near 1 .

As was seen in Example 6, no simple arc can fill the unit square $S \equiv [0, 1] \times [0, 1]$. By the same argument, *every simple arc in the plane is nowhere dense*. It would appear from this that a simple arc cannot occupy "very much" of S. In particular, it cannot occupy almost all of S, since if a simple arc A in S had measure equal to 1 it would be dense in S, and being closed it would be equal to S. However, it *is* possible for a simple arc A to have *positive* plane measure. Indeed, if ε is any number between 0 and 1, there exists a simple arc A whose plane measure is greater than $1 - \varepsilon$. We now outline a proof of this remarkable fact.

The general procedure will be to modify the construction given in Example 6 by cutting open "channels" between adjacent subsquares of S that do *not* correspond to adjacent subintervals of I. After the first stage the "subsquares" become *subquadrilaterals* which, in turn, are subdivided by lines joining opposite midpoints. Further open channels are cut out, and each closed quadrilateral is reduced to a sequence of *eight* subquadrilaterals. The first subdivision and the general scheme, where squares are used instead of general quadrilaterals, for simplicity, are shown in Figure 9. The second stage is shown in Figure 10. In both cases the channels deleted are indicated by shading. After n stages there are 8^n closed quadrilaterals, those numbered $8k - 7$ to $8k$ being subquadrilaterals of the quadrilateral numbered k at stage $n - 1$ ($k = 1, 2, \cdots, 8^{n-1}$). Furthermore , at each stage, two quadrilaterals are adjacent if and only if they bear

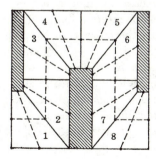
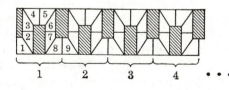

Figure 9

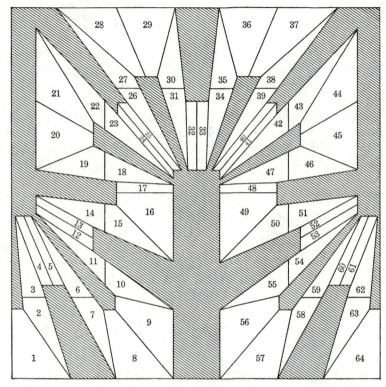

Figure 10

consecutive numbers, and hence correspond to adjacent subintervals of $I = [0, 1]$. It is not hard to show that the diameter of each quadrilateral is at most $\frac{3}{4}$ the diameter of the quadrilateral that contains it at the preceding stage. Consequently, any decreasing infinite sequence of quadrilaterals determines a unique point of intersection, and the mapping is well-defined, as in Example 6. Furthermore, this mapping is continuous for the same reasons that apply in Example 6, and is one-to-one because all irrelevant adjacencies have been removed. Finally, since the channels removed can be made of arbitrarily small area, their union can be made of arbitrarily small plane measure, and the simple arc remaining has plane measure arbitrarily near 1.

A second method of constructing a simple arc with positive plane

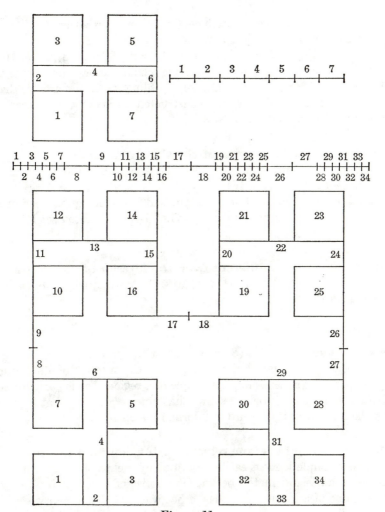

Figure 11

measure is indicated in Figure 11. This is somewhat simpler conceptually than the construction just described, but has the disadvantage that certain subintervals of [0, 1] are mapped onto sets of zero plane measure. The construction suggested in Figure 11 produces a simple arc containing $A \times A$, where A is a Cantor set. Since, for

$0 < \varepsilon \leqq 1$, we may chose A to have (linear) measure at least $\sqrt{1-\varepsilon}$, the arc in question has plane measure at least $1 - \varepsilon$.

The American mathematician W. F. Osgood (1864–1943), in 1903 (cf. [37]) constructed a simple arc having plane measure greater than $1 - \varepsilon$ by making use of a Cantor set A of linear measure greater than $\sqrt{1 - \varepsilon}$. The simple arc was constructed in such a way that it contains the product set $A \times A$.

11. A connected compact set that is not an arc.

The set S_2 of the second example of Example 3 is not an arc because it is not locally connected: If $N = \{(x, y) \mid (x, y) \in S_2, x^2 + y^2 < \frac{1}{2}\}$, there is no neighborhood of the origin that is a subneighborhood of N in which every two points can be joined by a connected set lying in N (cf. [17], p. 204).

12. A plane region different from the interior of its closure.

Let $S \equiv \{(x, y) \mid x^2 + y^2 < 1\} \setminus ([0, 1) \times \{0\})$, i.e., an open disk with a slit deleted. Then

$$\bar{S} = \{(x, y) \mid x^2 + y^2 \leqq 1\},$$

and the interior of $\bar{S}$ is $I(\bar{S}) = \{(x, y) \mid x^2 + y^2 < 1\}$.

Since every Jordan region is equal to the interior of its closure ([36], p. 477), this is a simple example of a region that is *not* a Jordan region. Example 14, below, shows that not *every* region that is equal to the interior of its closure is a Jordan region.

13. Three disjoint plane regions with a common frontier.

This example is most easily described by means of a story. A man lives on an island in the ocean. On the island are two fresh-water pools, one of cold and the other of hot water, and the man wishes to bring all three sources of water to within a convenient distance from any point of the island. He proceeds to dig channels, but always in such a way that the island remains homeomorphic to its original form. He starts by permitting the ocean to invade his island, coming within a distance of at most one foot of each point of his residual island (but not into contiguity with the fresh-water pools). He then extends the cold fresh-water domain in a similar fashion, and follows this by forming channels for the hot fresh-water, with the result that every point of the island thus remaining is within one foot of all three types of water supply. Unsatisfied with this result the islander repeats the triple process just described in order to have each type of water within a half-foot of each point remaining on the island. Again he is not satisfied, and refines the approximation to within a quarter-foot. He then extends this process to an infinite sequence of steps, each time halving the critical distance, and also the time of completion in order to finish in a finite length of time. If we assume that the original "island" is a compact disk with two inner disjoint open disks removed, and that the "ocean" is the open planar complement lying outside this disk, and that all extensions of the three original regions remain homeomorphic to their original forms, we obtain three final disjoint regions, R_1, R_2, and R_3, each being a union of the regions of an infinite sequence of regions. The final "island," similarly, is the intersection F of the islands of an infinite sequence, and is the common frontier of the three regions R_1, R_2, and R_3. Since the complement of F consists of three disjoint regions instead of two, no one of the regions R_1, R_2, and R_3 is a Jordan region. (For a discussion and proof of the Jordan curve theorem, see [33].) On the other hand, each of these three regions is equal to the interior of its closure as we shall see in the following Example 14.

The preceding construction may be carried out with any finite number of (indeed, with countably many) disjoint regions. If more than four regions are used we can thus produce a "map" in which all "countries" have a common frontier. This shows that the famous four-color problem requires careful formulation to avoid a trivial and negative solution. (Cf. [13].)

14. A non-Jordan region equal to the interior of its closure.

Let R be any one of the regions R_1, R_2, and R_3 defined in Example 13. As has been noted, R is not a Jordan region. On the other hand, R is equal to the interior of its closure, as we shall now demonstrate. In the first place, since $R \subset \bar{R}$ and R is open, $R = I(R) \subset I(\bar{R})$. We now wish to show the reverse inclusion: $I(\bar{R}) = I(R \cup F) \subset R$. If this were false, there would be a point p of F that is an interior point of $R \cup F$. But this means that there is a neighborhood N of p that lies in $R \cup F$ and therefore contains no points of either of the remaining two regions, in contradiction to the fact that every point of F is a limit point of each of the three regions R_1, R_2, and R_3.

15. A bounded plane region whose frontier has positive measure.

Let A be a Cantor set of positive measure in $[0, 1]$, and let

$$R \equiv ((0, 1) \times (-1, 1)) \setminus (A \times [0, 1)).$$

Then R is a region and

$$F(R) = (\{0\} \times [-1, 1]) \cup (\{1\} \times [-1, 1]) \cup (A \times [0, 1))$$

$$\cup ((0, 1) \times \{1\}) \cup ((0, 1) \times \{-1\}),$$

whence $\mu(F(R)) = \mu(A) > 0$. Clearly R is not a Jordan region $(I(\bar{R}) \neq R)$. (Cf. [14], p. 292.) (See Example 4, Chapter 11 for a *Jordan* region having a frontier with positive plane measure.)

16. A simple arc of infinite length.

First example: Let

$$f(x) \equiv \begin{cases} 0 & \text{if} \quad x = 1/n, \quad n \in \mathfrak{N}, \quad n \text{ odd}, \\ 1/n & \text{if} \quad x = 1/n, \quad n \in \mathfrak{N}, \quad n \text{ even}, \end{cases}$$

let $f(0) \equiv 0$, and let $f(x)$ be linear in each interval $[1/(n + 1), 1/n]$, $n \in \mathfrak{N}$. Then the graph of $f(x)$, for $x \in [0, 1]$, is a simple arc of infinite length because of the divergence of the harmonic series.

Second example: Let $f(x) \equiv x \sin(1/x)$ for $x \in (0, 1]$ and $f(0) \equiv 0$. The graph of $f(x)$, for $x \in [0, 1]$, is again a simple arc of infinite length for the same reason as in the first example. The lengths of inscribed polygonal arcs dominate sums of heights of individual arches of the graph of $f(x)$, and these sums have the form $\sum 2/(2n - 1)\pi$.

In contrast to the two preceding examples, the graph of the function defined $f(x) \equiv x^2 \sin(1/x)$ for $x \in (0, 1]$ and $f(0) \equiv 0$ is of finite length for $x \in [0, 1]$, since it is differentiable and its derivative is bounded there. (Cf. [36], p. 353 (Exs. 24 and 27), p. 176 (Theorem II).)

17. A simple arc of infinite length and having a tangent line at every point.

If $f(x) \equiv x^2 \sin(1/x^2)$ for $x \in (0, 1]$ and $f(0) \equiv 0$, the graph of

$f(x)$ for $x \in [0, 1]$ is a simple arc of infinite length for reasons given for the second example of Example 16. The graph of $f(x)$ has a tangent line at every point since $f(x)$ is everywhere differentiable.

18. A simple arc that is of infinite length between every pair of distinct points on the arc.

First example: Let $f(t) = (x(t), y(t))$ be any space-filling arc mapping $[0, 1]$ onto $[0, 1] \times [0, 1]$ and possessing the additional property that f maps every nondegenerate interval of $[0, 1]$ onto a set having a nonempty (two-dimensional) interior. (The mapping of Example 6 has this property.) Then the graph of either $x(t)$ or $y(t)$, for $t \in [0, 1]$, has the stated properties. To see that the graph of $x(t)$ for $a \leqq t \leqq b$ has infinite length, for example, we may use the fact (cf. Example 9) that $x(t)$ assumes at least two of its values uncountably many times each.

Second example: Let f be a mapping of the type described in Example 10 and such that every nondegenerate subinterval of $[0, 1]$ is mapped onto a set of positive plane measure. Then f has the properties specified above, since any rectifiable simple arc has plane measure zero (cf. [36], p. 436).

Third example: The graph of any function that is everywhere continuous and nowhere differentiable on a closed interval (cf. Example 8, Chapter 3) has the desired properties since if this graph were rectifiable the function would be of bounded variation, and every function of bounded variation is differentiable almost everywhere. (Cf. [16].)

Fourth example: Cf. [14], p. 190.

19. A smooth curve C containing a point P that is never the nearest point of C to any point on the concave side of C.

Let the curve C be the graph of $y^3 = x^4$, which is everywhere concave up, and let $P \equiv (0, 0)$. If (a, b) is a point lying above C and if $a \neq 0$, then clearly $(a, a^{4/3})$ is nearer (a, b) than $(0, 0)$ is. If b is an arbitrary number greater than or equal to 1, then the point $(\frac{1}{8}, \frac{1}{16})$ is nearer $(0, b)$ than $(0, 0)$ is. Finally, if b is an arbitrary positive number less than 1, then the point (b^3, b^4) is nearer $(0, b)$ than $(0, 0)$ is. The idea is that the origin is a point of infinite curvature (zero radius of curvature) of C. (Cf. [34], p. 258 (Ex. 16).)

20. A subset A of the unit square $S = [0, 1] \times [0, 1]$ that is dense in S and such that every vertical or horizontal line that meets S meets A in exactly one point.

What we are seeking is a one-to-one correspondence f with domain and range $[0, 1]$ and with a graph dense in S. We start by defining $f(x)$ for $x \in (0, 1] \cap \mathbb{Q}$, in stages. Let the points of $B \equiv ((0, 1] \cap \mathbb{Q}) \times ((0, 1] \cap \mathbb{Q})$ be arranged in a sequence: (x_1, y_1), (x_2, y_2), $\cdots$. We define $f(x_1) \equiv y_1$ for the zero stage. For stage one we partition B into four disjoint parts by vertical and horizontal bisecting lines, $((0, \frac{1}{2}] \cap \mathbb{Q}) \times ((0, \frac{1}{2}] \cap \mathbb{Q})$, $((0, \frac{1}{2}] \cap \mathbb{Q}) \times ((\frac{1}{2}, 1] \cap \mathbb{Q})$, $\cdots$, and denote these parts in any order: B_{11}, B_{12}, B_{13}, B_{14}. Denote by (x_{11}, y_{11}) the first point of the sequence $\{(x_n, y_n)\}$ that belongs to B_{11} and is such that neither $x_{11} = x_1$ nor $y_{11} = y_1$, and let $f(x_{11}) \equiv y_{11}$. Denote by (x_{12}, y_{12}) the first point of the sequence $\{(x_n, y_n)\}$ that belongs to B_{12} and is such that x_{12} is different from both x_1 and x_{11} and y_{12} is different from both y_1 and y_{11}, and let $f(x_{12}) \equiv y_{12}$. After $f(x_{13})$ is defined similarly to be equal to y_{13}, we denote by (x_{14}, y_{14}) the first point of $\{(x_n, y_n)\}$ that belongs to B_{14} and is such that x_{14} is different from x_1, x_{11}, x_{12}, and x_{13}, and y_{14} is different from y_1, y_{11}, y_{12}, and y_{13}, and define $f(x_{14}) \equiv y_{14}$. This completes stage one. Stage two is similar, with B partitioned into sixteen $= 4^2$ parts by further vertical and horizontal bisections, B_{21}, B_{22}, $\cdots$, $B_{2,4^2}$. For each of these parts in turn we define f at a rational point not yet in its domain and having as value a rational point not yet in its range. If this procedure is indefinitely continued, with B partitioned into 4^n congruent parts at stage n, a function f having the specified properties for $(0, 1] \cap \mathbb{Q}$ is obtained. Finally, we extend the domain and range of f to $[0, 1]$ by defining $f(x) \equiv x$ for $x \in [0, 1] \setminus ((0, 1] \cap \mathbb{Q})$.

21. A nonmeasurable plane set having at most two points in common with any line.

This example, due to W. Sierpinski [44], depends for its construction on the maximality principle, appearing in the form of the well-ordering theorem and also in the form of Zorn's lemma (cf. [16], [30], and [46]). We start by listing four statements having to do with cardinal and ordinal numbers:

(*i*) If $\mathfrak{a}$ is an infinite cardinal, then $\mathfrak{a}^2 = \mathfrak{a}$ (cf. [16] and [46]).

(*ii*) The cardinality $\mathfrak{f}$ of the set of closed sets of positive plane measure is c, the cardinal number of $\mathfrak{R}$. (Since the closed sets and their complements are in one-to-one correspondence, $\mathfrak{f} \leqq$ the cardinal number of the set of all open sets. Since every open set is a countable union of open disks having rational radii and centers with rational coordinates, we see that $\mathfrak{f} \leqq c$. Since the closed disks centered at the origin constitute a set of cardinality c, we see that $\mathfrak{f} \geqq c$.)

(*iii*) Let Ψ denote the first ordinal number corresponding to the cardinality c (cf. Example 10, Chapter 12). Then $\{\alpha \mid \alpha < \Psi\}$ has cardinality c.

(*iv*) If E is a linear measurable set of positive linear measure, then the cardinality of E is c. (E contains a closed linear set F of positive linear measure. F is the union of a countable set (possibly empty) and a (necessarily nonempty) perfect set. (Cf. [20] and [45].))

Let $\alpha \rightarrow F_\alpha$ be a one-to-one mapping of the set $\{\alpha \mid \alpha < \Psi\}$ onto the set of all closed sets of positive plane measure. Let $\mathfrak{F}$ be the family of all functions $p(\alpha)$ whose domains have the form $[1, \beta)$ for some $\beta \leqq \Psi$, whose ranges are subsets of the plane, and are such that

(a) $p(\alpha) \in F_\alpha$ for every $\alpha \in$ domain of $p(\alpha)$,

(b) no three points in the range of $p(\alpha)$ are collinear.

Let $\mathcal{G}$ be the set of all ranges of the functions in $\mathfrak{F}$, and let $\mathcal{G}$ be partially ordered by inclusion. Then by Zorn's lemma (cf. [16], [30], and [46]) there exists a maximal set $E \in \mathcal{G}$, which is the range of a function $q(\alpha)$ of the set $\mathfrak{F}$. If the domain of $q(\alpha)$ is $[1, \beta)$, we shall now show that $\beta = \Psi$ by assuming the contrary, $\beta < \Psi$, and obtaining a contradiction. If $\mathfrak{b}$ is the cardinal number corresponding to β, then $\mathfrak{b} \leqq \mathfrak{b}^2 < c$ (the strict inequality $\mathfrak{b} < \mathfrak{b}^2$ holding iff $1 < \mathfrak{b}$ and $\mathfrak{b}$ is finite). This means that the cardinality of the set of all directions determined by pairs of points in the range E of $q(\alpha)$ is less than c, and therefore that there exists a direction θ different from all directions determined by pairs of points in E. Then some line L in the direction θ must meet the set F_β in a set of positive linear measure (by the Fubini theorem). Since this latter set has cardinality c there is a point $p_\beta \in F_\beta$ such that p_β is collinear with no pair of points in the range of $q(\alpha)$. We now extend the function $q(\alpha)$ so that its domain is $[1, \beta] = [1, \beta + 1)$, and so that $q(\beta) = p_\beta$. Then this extended function $q(\alpha)$ has both properties (a) and (b) and its range is strictly greater than the *maximal* member E of $\mathcal{G}$. This is the desired contradiction, and therefore $\beta = \Psi$, the domain of the function $q(\alpha)$

consists of *all* α less than Ψ, and the range E of $q(\alpha)$ contains a point p_α from *every* closed plane set F_α of positive plane measure.

We now show that the set E is nonmeasurable by assuming the contrary and obtaining a contradiction. Indeed, if E is measurable, then so is its complement E', and since E' contains no closed plane set of positive plane measure, E' must have measure zero. On the other hand, since every line in the plane meets E in at most two points, E must also have measure zero (by the Fubini theorem). Therefore the entire plane, being the union of the two sets E and E' of measure zero, must also be of measure zero, and we have the desired contradiction.

We note too that if S is any set of positive plane measure, then $S \cap E$ is nonmeasurable. Otherwise the Fubini theorem implies that $\mu(S \cap E) = 0$, whence $\mu(S \setminus E) > 0$. Thus $S \setminus E$ contains some closed set F of positive plane measure. Since $F \cap E = \emptyset$, there is a contradiction of the basic property of E: E meets every closed set of positive plane measure.

S. Mazurkiewicz [31] constructed a plane set E meeting each line of the plane in *precisely* two points. However, such a set E may be measurable and indeed is then of measure zero. The reason for this is the form of the construction which depends only upon the existence of a set E_1 in the plane such that E_1 meets every line in a set of cardinality c. The set E is then formed as a subset of E_1.

However, sets enjoying the property of E_1 may have plane measure zero. For example, let C be the Cantor set on $[0, 1]$ and let

$$E_1 = (\Re \times C) \cup (C \times \Re).$$

Then clearly each line meets E_1 in a set of cardinality c and yet E_1 is a (closed) set of plane measure zero.

In [3] the construction of a "Mazurkiewicz set" is given in answer to a problem posed in that journal.

F. Galvin has shown the following: If $1 < n \leqq \aleph_0$, where $\aleph_0$ is the cardinality of $\Re$, there is a nonmeasurable set S in the plane such that the intersection of S with any line consists of precisely n points.

22. A nonnegative function of two variables $f(x, y)$ such that

$$\int_0^1 \int_0^1 f(x, y) \, dx \, dy = \int_0^1 \int_0^1 f(x, y) \, dy \, dx = 0$$

and such that $\iint_S f(x, y) \, dA$, **where** $S = [0, 1] \times [0, 1]$, **does not exist.**

We shall give two examples, one in which Riemann integration is used and one in which Lebesgue integration is used.

First example: Let f be the characteristic function of the set of Example 20. Then for every $y \in [0, 1]$, $\int_0^1 f(x, y) \, dx = 0$, and for every $x \in [0, 1]$, $\int_0^1 f(x, y) \, dy = 0$, the integrals being those of Riemann. However, the double Riemann integral over S fails to exist, since for the function f the upper and lower Riemann integrals are equal to 1 and 0, respectively.

Second example: Let f be the characteristic function of the set of Example 21. Then the iterated integrals are again both equal to zero, where the integration is that of either Riemann or Lebesgue, while the function f is not measurable on S, and hence has no double Lebesgue integral there.

23. A real-valued function of one real variable whose graph is a nonmeasurable plane set.

Let $f(x)$ be defined as follows, for $x \in \Re$:

$$f(x) \equiv \begin{cases} \max\{y \mid (x, y) \in E\} & \text{if} \quad \{y \mid (x, y) \in E\} \neq \emptyset, \\ 0 & \text{if} \quad \{y \mid (x, y) \in E\} = \emptyset, \end{cases}$$

where E is the set of Example 21. Let $E_1 \equiv \{(x, f(x)) \mid x \in \Re\} \cap E$, $E_2 \equiv E \setminus E_1$. Then either E_1 or E_2 (or both) must be nonmeasurable since their union is E. If E_1 is nonmeasurable, then

$$F \equiv \{(x, f(x)) \mid x \in \Re\},$$

the graph of f, is the union of E_1 and a subset of the x axis; hence, since the latter has plane measure zero, F is nonmeasurable. If E_2 is nonmeasurable, let $g(x)$ be defined, for $x \in \Re$:

$$g(x) \equiv \begin{cases} \min\{y \mid (x, y) \in E)\} & \text{if} \quad \{y \mid (x, y) \in E\} \\ & \text{consists of two distinct points,} \\ 0 & \text{otherwise.} \end{cases}$$

Then $G \equiv \{(x, g(x)) \mid x \in \Re\}$, the graph of g, is the union of E_2 and a subset of the x axis, and hence nonmeasurable. There must exist, then, in one way or the other, a function whose graph is a nonmeasurable plane set.

24. A connected set that becomes totally disconnected upon the removal of a single point.

We give only a sketch. For details see [25].

Let C be the Cantor set of Example 1, Chapter 8, let B be the subset of C consisting of all endpoints of the open intervals that were deleted from $[0, 1]$ in the construction of C, and let $E \equiv C \setminus B$ (cf. Example 24, Chapter 8). For every $x \in C$ let $L(x)$ be the closed segment joining the points $(x, 0)$ and $(1, 1)$ in the plane. If $x \in B$ let $S(x)$ consist of those points of $L(x)$ whose ordinates are irrational, and if $x \in E$ let $S(x)$ consist of those points of $L(x)$ whose ordinates are rational. Then $S \equiv \bigcup_{x \in C} S(x)$ is a set having the required properties.

The connectedness of S is proved by means of arguments involving sets of the first and second categories, and we shall omit the discussion. If $S_0 \equiv S \setminus \{(1, 1)\}$, then S_0 is totally disconnected. For if $E \subset S_0$, if E contains more than one point, and if E is a subset of any $S(x)$, $x \in C$, then E is clearly not connected. On the other hand, if p and q are two points of S_0 on two *distinct* intervals $L(x)$ and $L(y)$, where x and $y \in C$ and $x < y$, there is in the complement of S_0 a straight line through $(1, 1)$ that separates p and q, namely the straight line passing through $(1, 1)$ and any $(a, 0)$ where $x < a < y$ and $a \notin C$.

Chapter 11
Area

Introduction

The concept of *area* is based on that of the Riemann double integral. A bounded plane set S is said to **have area** iff its characteristic function χ_S is (Riemann) integrable over a closed rectangle R containing S and such that the sides of R are parallel to the coordinate axes. If S has area, its **area** $A(S)$ is equal to the double integral of χ_S over R:

$$A(S) \equiv \iint_R \chi_S \, dA.$$

These definitions are meaningful in the sense that the concepts of *having area* and of *area* are independent of the containing rectangle R. If R is subdivided into closed subrectangles by means of a net $\mathfrak{N}$ of lines parallel to the sides of R, then some of these subrectangles may be subsets of S, and some may be subsets of the complement S' of S. For any such net $\mathfrak{N}$, let $a(\mathfrak{N})$ be the sum of the areas of all subrectangles that are subsets of S ($a(\mathfrak{N}) = 0$ in case there are no such subrectangles), and let $A(\mathfrak{N})$ be the sum of the areas of all subrectangles that are *not* subsets of S' (that is, that intersect S nonvacuously). The **inner area** and **outer area** of S, denoted $\underline{A}(S)$ and $\bar{A}(S)$, respectively, are defined as the supremum of $a(\mathfrak{N})$ and the infimum of $A(\mathfrak{N})$, respectively, for all nets $\mathfrak{N}$ of lines parallel to the sides of R:

$$\underline{A}(S) \equiv \sup a(\mathfrak{N}), \qquad \bar{A}(S) \equiv \inf A(\mathfrak{N}).$$

Again, these definitions are independent of R. A bounded set S has area iff $\underline{A}(S) = \bar{A}(S)$, and in case of equality, $A(S) = \underline{A}(S) = \bar{A}(S)$.

A necessary and sufficient condition for a bounded set S to have area is that its frontier $F(S)$ have zero area, or equivalently, that $F(S)$ have zero outer area. Since for any bounded set S, $F(S)$ is a compact set (and hence measurable as a plane set), and since for compact sets outer area and outer plane (Lebesgue) measure are identical, a bounded set has area iff its frontier has plane measure zero.

The preceding statements concerning area apply in similar fashion to *volume*, for sets in three-dimensional Euclidean space. A generalization of *area* and *volume* that applies to Euclidean spaces of any number of dimensions — and, indeed, to much more general spaces — is Jordan content. (Cf. [36], p. 431.) Lebesgue measure is a generalization of Jordan content in the sense that every set that has content is measurable, and its content and measure are equal. The principal advantages of Lebesgue measure over Jordan content lie in the broader applicability of measure to limiting processes. For an elementary treatment of plane area and volume, including proofs of many of the preceding statements, cf. [36], pp. 431–465.

Examples 7 and 8 of this chapter concern *surface area*. For a discussion of this subject see the references given in connection with these two examples.

1. A bounded plane set without area.

The set $S \equiv (\mathbb{Q} \cap [0, 1]) \times (\mathbb{Q} \cap [0, 1])$ of points in the unit square both of whose coordinates are rational is without area since its frontier $F(S)$ does not have zero area. (The set $F(S)$ is the unit square itself and hence has area equal to 1.) The outer area of S is 1 and its inner area is 0.

2. A compact plane set without area.

Let A be a Cantor set of positive measure ε (Example 4, Chapter 8), and let $S \equiv A \times [0, 1]$. Then $F(S) = S$, and the plane measure of $F(S)$ is equal to the linear measure ε of A. Since $F(S) = S$ is a compact set its outer area is equal to its measure, and is thus positive. Therefore S is without area. The outer area of S is equal to ε and its inner area is 0.

3. A bounded plane region without area.

The region R of Example 15, Chapter 10 is bounded and without area.

4. A bounded plane Jordan region without area.

Let ε be a positive number less than 1, and let A be a simple arc with parametrization $f(t)$, $0 \leqq t \leqq 1$, lying in the unit square $[0, 1] \times [0, 1]$, and of plane measure greater than $1 - \frac{1}{2}\varepsilon$ (cf. Example 10, Chapter 10). Let C be the simple closed curve formed by the union of A and the three segments $\{0\} \times [-\frac{1}{2}\varepsilon, 0], \{1\} \times [-\frac{1}{2}\varepsilon, 0]$, and $[0, 1] \times \{-\frac{1}{2}\varepsilon\}$, and let R be the bounded region having C as its frontier. Then R is a Jordan region and its frontier has outer area greater than $1 - \frac{1}{2}\varepsilon > 1 - \varepsilon > 0$.

5. A simple closed curve whose plane measure is greater than that of the bounded region that it encloses.

If C and R are the curve and region defined in Example 4 and if μ is plane Lebesgue measure, then

$$\mu(R \cup C) = \mu(R) + \mu(C) \leqq 1 + \tfrac{1}{2}\varepsilon.$$

Therefore, since $\mu(C) > 1 - \frac{1}{2}\varepsilon$, it follows that

$$\mu(R) < \varepsilon.$$

The measure of R is less than that of C whenever $\varepsilon < \frac{2}{3}$. Simultaneously, the measures of R and C can be made arbitrarily near 0 and 1, respectively.

6. Two functions ϕ and ψ defined on $[0, 1]$ and such that

(a) $\phi(x) < \psi(x)$ for $x \in [0, 1]$,
(b) $\int_0^1 [\psi(x) - \phi(x)] \, dx$ exists and is equal to 1,
(c) $S \equiv \{(x, y) \mid 0 \leqq x \leqq 1, \phi(x) < y < \psi(x)\}$ is without area.

Let $\phi(x)$ be the characteristic function of $\mathbb{Q} \cap [0, 1]$ and let $\psi(x) \equiv \phi(x) + 1$. Then (a) and (b) are clearly satisfied, while $F(S)$ is the closed rectangle $[0, 1] \times [0, 2]$ of positive area, whence S is without area. The outer area of S is 2 and its inner area is 0.

This example is of interest in connection with **Cavalieri's Principle,** which states that if every plane Π parallel to a given plane Π_0 intersects two three-dimensional sets W_1 and W_2 in plane sections of equal area, then W_1 and W_2 have equal volume (cf. [18]). A two-dimensional analogue states that if every line L parallel to a given line L_0 intersects two plane sets S_1 and S_2 in segments of equal length, then S_1 and S_2 have equal area. The present example shows that unless S_1 and S_2 are assumed to have area, this statement is false. (The sets S_1 and S_2 can be taken to be the set S of (c) and the closed square $[0, 1] \times [3, 4]$, respectively, with the family of parallel lines being the family of all vertical lines.) Construction of a three-dimensional counterexample to Cavalieri's Principle is left as an exercise for the reader.

7. A means of assigning an arbitrarily large finite or infinite area to the lateral surface of a right circular cylinder.
Let S be the right circular cylinder

$$S \equiv \{(x, y, z) \mid x^2 + y^2 = 1, \quad 0 \leqq z \leqq 1\}$$

of base radius 1 and altitude 1, and for each positive integer m let the $2m + 1$ circles C_{km} be defined, for $k = 0, 1, \cdots, 2m$:

$$C_{km} \equiv S \cap \{(x, y, z) \mid z = k/2m\}.$$

On each of these $2m + 1$ circles let the n equally spaced points P_{kmj} be defined for each positive integer n and for $j = 0, 1, \cdots, n - 1$:

$$P_{kmj} \equiv \begin{cases} \left(\cos \dfrac{2j\pi}{n}, \sin \dfrac{2j\pi}{n}, \dfrac{k}{2m} \right) & \text{if } k \text{ is even,} \\[3mm] \left(\cos \dfrac{(2j+1)\pi}{n}, \sin \dfrac{(2j+1)\pi}{n}, \dfrac{k}{2m} \right) & \text{if } k \text{ is odd.} \end{cases}$$

For each circle C_{km} the points $P_{kmj}, j = 0, 1, \cdots, n$, are the vertices of a regular polygon of n sides. If $0 < k \leqq 2m$, each side of the polygon with vertices lying on the circle C_{km} lies above a vertex of the polygon in $C_{k-1,m}$ and thus determines a (plane) triangle in space. Similarly, if $0 \leqq k < 2m$, each side of the polygon in C_{km} lies below a vertex of the polygon in $C_{k+1,m}$ and thus determines a triangle. It is not difficult to see that there are a total of $4mn$ congruent space triangles formed in this way with vertices lying on the given cylinder, and a little trigonometry shows that the area of each of these triangles is $\sin (\pi/n) [(1/4m^2) + (1 - \cos (\pi/n))^2]^{1/2}$. The area of the polyhedron Π_{mn} inscribed in S is therefore obtained by multiplying this quantity by $4mn$. The result can be expressed

$$A(\Pi_{mn}) = 2\pi \frac{\sin(\pi/n)}{(\pi/n)} \sqrt{1 + 4m^2\left(1 - \cos\frac{\pi}{n}\right)^2}.$$

As m and $n \to +\infty$, the diameters of the triangles approach zero and thus, presumably, the areas of the inscribed polyhedra should approach a limit, and it is natural to expect that this limit should be the number $(2\pi 1)\cdot 1 = 2\pi$ given by the familiar formula $2\pi rh$ for the area of the lateral surface of a right circular cylinder, where r is the base radius and h the altitude. We shall see, however, that the result will depend on the *relative* rates at which m and n increase.

We observe first that as $n \to +\infty$ the factor preceding the radical in the formula for $A(\Pi_{mn})$ has the limit 2π, and since the radical itself is at least as great as 1, any limit that $A(\Pi_{mn})$ may have must be *at least* 2π. We concentrate our attention now on the quantity within the radical and, in fact, on the function

$$f(m, n) \equiv 2m\left(1 - \cos\frac{\pi}{n}\right) = \frac{\pi^2 m}{n^2} - \frac{2\pi^4 m}{4!\, n^4} + \frac{2\pi^6 m}{6!\, n^6} - \cdots.$$

We shall consider three cases:

(*i*) If $m = n$, then

$$f(n, n) = 2n\left(1 - \cos\frac{\pi}{n}\right) = \frac{\pi^2}{n} - \frac{2\pi^4}{4!\, n^3} + \cdots,$$

$\lim_{n\to+\infty} f(n, n) = 0$, and $\lim_{n\to+\infty} A(\Pi_{nn}) = 2\pi$.

(*ii*) If $m = [\alpha n^2]$, where the brackets indicate the bracket function of Chapter 2, and where $0 < \alpha < +\infty$, then

$$f([\alpha n^2], n) = 2[\alpha n^2]\left(1 - \cos\frac{\pi}{n}\right) = \frac{\pi^2[\alpha n^2]}{n^2} - \frac{2\pi^4[\alpha n^2]}{4!\, n^4} + \cdots,$$

$\lim_{n\to+\infty} f([\alpha n^2], n) = \alpha\pi^2$, and $\lim_{n\to+\infty} A(\Pi_{[\alpha n^2],n}) = 2\pi\sqrt{1 + \alpha^2\pi^4}$.

(*iii*) If $m = n^3$, then

$$f(n^3, n) = 2n^3\left(1 - \cos\frac{\pi}{n}\right) = \pi^2 n - \frac{2\pi^4}{4!\, n} + \cdots,$$

$\lim_{n\to+\infty} f(n^3, n) = +\infty$, and $\lim_{n\to+\infty} A(\Pi_{n^3,n}) = +\infty$.

We conclude that as m and $n \rightarrow +\infty$, any result, finite or infinite, that is at least equal to 2π can be obtained for the limit of $A(\Pi_{mn})$. Although, in general, $\lim_{m,n\rightarrow+\infty} A(\Pi_{mn})$ does not exist, we can at least say that the *limit inferior* exists and that

$$\lim_{m,n \rightarrow +\infty} A(\Pi_{mn}) = 2\pi.$$

The example just described is due to H. A. Schwarz (*Gesammelte Mathematische Abhandlungen*, Vol. 2, p. 309 (Berlin, Julius Springer, 1890.)). It serves to demonstrate that the concept of *surface area* is far more complicated than that of *arc length*. For a discussion of surface area and further references, see [40]. An elementary treatment of surface area is given in [34], pp. 610–635.

8. For two positive numbers ε and M, a surface S in three-dimensional space such that:

(a) **S is homeomorphic to the surface of a sphere,**
(b) **The surface area of S exists and is less than ε,**
(c) **The three-dimensional Lebesgue measure of S exists and is greater than M.**

This example is due to A. S. Besicovitch (cf. [7]). The ideas involved in this construction are somewhat similar to those involved in the construction of a simple arc of positive plane measure (Example 10, Chapter 10), but far more complicated and sophisticated. Since an ample discussion would require a definition of surface area as well as an intricate description of tubular connections among faces of cubes, we shall omit the particulars.

The following discussion points up some interesting aspects of Example 8.

a. There is an analogy between Example 8 and Example 5. In both cases, there is more in the sides of the container than in the interior of the container. However, the linear measure (length) of the bounding curve of Example 5 is infinite whereas the planar measure (surface area) of the bounding surface of Example 8 is finite and small.

b. The familiar relations between the volume of a cube and its surface area: volume $= \frac{1}{6} \cdot$ edge$\cdot$surface area, and between the volume of a sphere and its surface area: volume $= \frac{1}{3} \cdot$ radius$\cdot$surface area, lead one to feel that a closed surface of small area together with the three-dimensional region that it encloses should have small three-dimensional measure. Example 8 is a counterexample to this feeling.

c. A right cylindrical "can" of finite height and based on a non-rectifiable Jordan curve has finite volume (three-dimensional measure) and infinite surface area. (The can can be filled with paint, but its sides cannot be painted.) This example is a weak dual to Example 8.

9. A plane set of arbitrarily small plane measure within which the direction of a line segment of unit length can be reversed by means of a continuous motion.

This example was given in 1928 by A. S. Besicovitch as a solution to a problem posed in 1917 by S. Kakeya. (Cf. [5], [6], and [23], and for an expository discussion, [8].)

Chapter 12
Metric and Topological Spaces

Introduction

A **metric space** is an ordered pair (X, d), where X is a nonempty set and d a real-valued function in $X \times X$ such that

(i) d is **strictly positive:**

$$x \in X \Rightarrow d(x, x) = 0,$$

$$x \text{ and } y \in X, x \neq y \Rightarrow d(x, y) > 0;$$

(ii) the **triangle inequality** holds:

$$x, y, \text{ and } z \in X \Rightarrow d(x, z) \leqq d(y, x) + d(y, z).$$

An early consequence of (i) and (ii) is

(iii) d is **symmetric:**

$$x \text{ and } y \in X \Rightarrow d(x, y) = d(y, x).$$

The function d is called the **metric** for the metric space (X, d), and the number $d(x, y)$ is called the **distance** between the points x and y. If the metric is clear from context, the single letter X may be used to represent both a metric space and the set of its points.

A **topological space** is an ordered pair (X, Θ), where X is a nonempty set and Θ is a family of subsets of X such that

(i) $\emptyset \in \Theta$ and $X \in \Theta$,

(ii) Θ is closed with respect to finite intersections:

$$O_1, \cdots, O_n \in \Theta \Rightarrow O_1 \cap \cdots \cap O_n \in \Theta,$$

where n is an arbitrary positive integer;

(iii) Θ is closed with respect to arbitrary unions:

$$(\lambda \in \Lambda \Rightarrow O_\lambda \in \Theta) \Rightarrow \bigcup_{\lambda \in \Lambda} O_\lambda \in \Theta,$$

where Λ is an arbitrary nonempty **index set**.

The family Θ is called **the topology** of the topological space (X, Θ) and its members are called **open sets**. The family Θ is also called **a topology** for the set X. If the family of open sets is clear from context, the single letter X may be used to represent both a topological space and the set of its points. By finite induction, condition (ii) is equivalent to the same for the special case $n = 2$. A topological space $(Y, \mathfrak{I})$ is a **subspace** of a topological space (X, Θ) iff $Y \subset X$ and $\mathfrak{I} = \{Y \cap O \mid O \in \Theta\}$; in this case the topology $\mathfrak{I}$ is said to be **induced** or **inherited** from Θ.

A set is **closed** iff its complement is open. An **open covering** of a set A is a class of open sets whose union contains A. A set C is **compact** iff every open covering of C contains a finite subcovering. A **Hausdorff space** is a topological space such that whenever x and y are two distinct points of the space there exist two disjoint open sets of which one contains x and one y. In any Hausdorff space every compact set is closed. A point p is a **limit point** of a set A iff every open set containing p contains at least one point of $A \setminus \{p\}$. The **closure** $\bar{A}$ of a set A is the intersection of all closed sets containing A, and consists of all points that are either members of A or limit points of A. The closure of any set A is closed. A set A is closed iff it is equal to its closure: $A = \bar{A}$. A **locally compact space** is a topological space such that every point is contained in an open set whose closure is compact.

A **base** for the topology of a topological space (X, Θ) is a subfamily $\mathcal{G}$ of Θ having the property that every nonempty member of Θ is the union of a collection of members of $\mathcal{G}$. A **neighborhood system** for a topological space (X, Θ) is a collection $\mathfrak{M}$ of ordered pairs (x, N) such that $x \in N$ for every $(x, N) \in \mathfrak{M}$, and the collection of all N such that $(x, N) \in \mathfrak{M}$ is a base for Θ. An example of a neighborhood system is the set of all (x, A) such that $x \in A$ and $A \in \mathcal{G}$, where $\mathcal{G}$ is a base for (X, Θ). If $\mathcal{G}$ is any nonempty family of subsets of a set X, then $\mathcal{G}$ is a base for some topology Θ for X iff (i) X is the union of the members of $\mathcal{G}$ and (*ii*) whenever G_1 and G_2 are members of $\mathcal{G}$ with a nonempty intersection, and $x \in G_1 \cap G_2$, then there exists a member G of $\mathcal{G}$ such that $x \in G \subset G_1 \cap G_2$. If (*i*) and (*ii*) hold, the topology Θ **generated** by $\mathcal{G}$ consists of the sets that are unions of members of $\mathcal{G}$. A sequence $\{x_n\}$ in a topological space **converges** to a point x, and x is a **limit** of the sequence $\{x_n\}$, iff

$$\forall \text{ open set } O \text{ containing } x \,, \exists \, m \in \mathfrak{N} \ni$$

$$n \in \mathfrak{N}, n > m \Rightarrow x_n \in O.$$

In any Hausdorff space limits of convergent sequences are unique.

If Θ_1 and Θ_2 are two topologies for the same set X, and if $\Theta_1 \subset \Theta_2$, then Θ_1 is said to be **weaker** than Θ_2 and Θ_2 is said to be **stronger** than Θ_1. The weakest of all topologies on X is the **trivial topology** $\Theta \equiv \{\emptyset, X\}$, and the strongest of all topologies is the **discrete topology** $\Theta \equiv 2^X$ consisting of *all* subsets of X.

If (X, d) is a metric space and if $x \in X$, then a **neighborhood**, or **spherical neighborhood**, of x is a set of the form

$$\{y \mid y \in X, d(x, y) < \varepsilon\},$$

where $\varepsilon > 0$ (x is called the **center** and ε the **radius** of this spherical neighborhood). A spherical neighborhood is sometimes called an **open ball**. The set of all spherical neighborhoods for any metric space satisfies the two conditions necessary for the generation of a topology, and for this topology, called **the topology** of the metric space, the set of all ordered pairs (x, N), where N is a spherical neighborhood of x, is a neighborhood system. A topological space (X, Θ) is **metrizable** iff there exists a metric d for X such that Θ is the topology of the metric space (X, d). A **closed ball** in a metric space (X, d) is a set of the form

$$\{y \mid y \in X, d(x, y) \leqq \varepsilon\},$$

where $x \in X$ and $\varepsilon > 0$, and is a closed set (x is called the **center** and ε the **radius**). The single unmodified word **ball** should be construed as synonymous with *closed ball*. A set in a metric space is **bounded** iff it is a subset of some ball. If the space (X, d) is bounded, then d is called a **bounded metric**. If the metric spaces (X, d) and (X, d^*) have the same topology, then d and d^* are called **equivalent**

metrics. If (X, d) is any metric space, then d^*, defined

$$d^*(x, y) \equiv \frac{d(x, y)}{1 + d(x, y)},$$

is a bounded metric equivalent to d; that is, *every metrizable space can be metrized by a bounded metric.* In any finite-dimensional Euclidean space with the standard Euclidean metric a set is compact iff it is closed and bounded. A sequence $\{x_n\}$ of points in a metric space (X, d) is a **Cauchy sequence** iff

$$\forall\, \varepsilon > 0\ \exists\ K \in \mathfrak{N}\ \ni$$

$$\left.\begin{array}{l} m \text{ and } n \in \mathfrak{N}, \\ m > K, \text{ and } n > K \end{array}\right\} \Rightarrow d(x_m, x_n) < \varepsilon.$$

A metric space is **complete** iff every Cauchy sequence of points in the space converges (to a point of the space). A metric space that is not complete is **incomplete.** Such concepts as *connected set, totally disconnected set,* and *perfect set* are defined exactly as in Euclidean spaces (cf. the Introduction, Chapter 10).

A topological space satisfies the **second axiom of countability** iff there exists a countable base for its topology. A set in a topological space is **dense** iff its closure is equal to the space. A topological space is **separable** iff it contains a countable dense set. A metrizable space satisfies the second axiom of countability iff it is separable.

If (X, Θ) and $(Y, \mathfrak{I})$ are topological spaces, and if f is a function on X into Y, then f is **continuous** iff $B \in \mathfrak{I} \Rightarrow f^{-1}(B) \in \Theta$; f is **open** iff $A \in \Theta \Rightarrow f(A) \in \mathfrak{I}$; f is **closed** iff $A' \in \Theta \Rightarrow (f(A))' \in \mathfrak{I}$. If f is a mapping of a topological space X onto a topological space Y, then f is a **topological mapping,** or a **homeomorphism** iff f is a one-to-one correspondence, and both f and f^{-1} are continuous.

Let V be an additive group, with members $x, y, z, \cdots$, and let F be a field, with members $\lambda, \mu, \nu, \cdots$. Then V is a **vector space** or **linear space** over F iff there exists a function $(\lambda, x) \to \lambda x$ on $F \times V$ into V such that for all λ and μ in F and x and y in V:

(*i*) $\lambda(x + y) = \lambda x + \lambda y$,

(*ii*) $(\lambda + \mu)x = \lambda x + \mu x$,

(*iii*) $\lambda(\mu x) = (\lambda \mu)x$,

(*iv*) $1x = x$.

The points of a vector space are also called **vectors**. If F is either the field $\mathfrak{R}$ of real numbers or the field $\mathfrak{C}$ of complex numbers, V is called a **normed vector space** over F iff there exists a real-valued **norm** $\|\ \ \|$ with the properties, for all x and y in V and λ in F:

(v) $\| x \| \geqq 0; \| x \| = 0$ iff $x = 0$,

(vi) $\| x + y \| \leq \| x \| + \| y \|$,

(vii) $\| \lambda x \| = | \lambda | \cdot \| x \|$.

Any normed vector space is a metric space with metric $d(x, y) \equiv \| x - y \|$. A **Banach space** is a complete normed vector space.

For further information on topological spaces and mappings see [11], [17], [20], [24], [27], [45], and [50]. For vector spaces in general, see [22]. For Banach spaces see [4] and [29].

1. A decreasing sequence of nonempty closed and bounded sets with empty intersection.

In the space $\mathfrak{R}$ with the bounded metric $d(x, y) \equiv \dfrac{| x - y |}{1 + | x - y |}$, let $F_n \equiv [n, +\infty), n = 1, 2, \cdots$. Then each F_n is closed and bounded, and $\bigcap_{n=1}^{+\infty} F_n = \varnothing$.

Since an empty intersection is impossible if the nonempty sets are *compact*, this example is impossible in any finite-dimensional Euclidean space with the standard Euclidean metric.

2. An incomplete metric space with the discrete topology.

The space $(\mathfrak{N}, d)$ of natural numbers with the metric $d(m, n) \equiv | m - n | / mn$ has the discrete topology since every one-point set is open, but the sequence $\{n\}$ is a nonconvergent Cauchy sequence.

This example demonstrates that *completeness is not a topological property*, since the space $\mathfrak{N}$ with the standard metric is both complete and discrete. In other words, it is possible for two metric spaces to be homeomorphic even though one is complete and one is not. Another example of two such spaces consists of the two homeomorphic intervals $(-\infty, +\infty)$ and $(0, 1)$ of which only the first is complete in the standard metric of $\mathfrak{R}$.

3. A decreasing sequence of nonempty closed balls in a complete metric space with empty intersection.

In the space $(\mathfrak{N}, d)$ of natural numbers with the metric

$$d(m, n) \equiv \begin{cases} 1 + \dfrac{1}{m + n} & \text{if } \quad m \neq n, \\[2mm] 0 & \text{if } \quad m = n, \end{cases}$$

let

$$B_n \equiv \{m \mid d(m, n) \leqq 1 + (1/2n)\} = \{n, n + 1, \cdots\},$$

for $n = 1, 2, \cdots$. Then $\{B_n\}$ satisfies the stipulated conditions, and the space is complete since every Cauchy sequence is "ultimately constant."

Trivial examples are possible if completeness is omitted — for example, $\{y \mid |(1/n) - y| \leqq (1/n)\}$ in the space $\mathcal{O}$ of positive numbers with the standard $\mathfrak{R}$ metric. On the other hand, the present example is impossible if the complete metric space is a Banach space (cf. [15]).

This example (cf. [45] (Sierpinski)) is of interest in connection with Baire's category theorem (cf. Example 7, Chapter 8, and [1], [4], [20], and [27]), which states that *every complete metric space is of the second category* or, equivalently, that any countable intersection of dense open sets in a complete metric space is dense. The proof involves the construction of a decreasing sequence of closed balls, with radii tending toward zero, and having *as a consequence* a nonempty intersection. We see, then, that if the balls get small they must have a point in common, whereas if they do *not* get small they may have nothing in common!

4. Open and closed balls, O and B, respectively, of the same center and radius and such that $B \neq \bar{O}$.

Let X be any set consisting of more than one point, and let (X, d) be the metric space with

$$d(x, y) \equiv \begin{cases} 1 & \text{if } \quad x \neq y, \\ 0 & \text{if } \quad x = y. \end{cases}$$

Let x be any point of X, and let O and B be the open and closed balls, respectively, with center x and radius 1. Then

$$O = \{x\}, \qquad B = X,$$

and since the topology is discrete, $\bar{O} = O \neq B$.

This example is impossible in any normed vector space. (Proof of this fact is left as an exercise.)

5. Closed balls B_1 and B_2, of radii r_1 and r_2, respectively, such that $B_1 \subset B_2$ and $r_1 > r_2$.

Let (X, d) be the metric space consisting of all points (x, y) in the closed disk in the Euclidean plane, defined:

$$X \equiv \{(x, y) \mid x^2 + y^2 \leq 9\},$$

with the standard Euclidean metric. Let $B_2 \equiv X$, and let

$$B_1 \equiv B_2 \cap \{(x, y) \mid (x - 2)^2 + y^2 \leq 16\}.$$

Then $B_1 \subset B_2$, and $r_1 = 4 > r_2 = 3$.

This example is impossible in any normed vector space since the radius of any ball is half its diameter. (Proof is left as an exercise.)

6. A topological space X and a subset Y such that the limit points of Y do not form a closed set.

Let X be any set consisting of more than one point, and let the topology of X be the trivial topology $\mathcal{O} \equiv \{\emptyset, X\}$. If y is an arbitrary member of X, let $Y \equiv \{y\}$. Then the limit points of Y are all points of X *except* for y itself. That is, the set of limit points is $X \setminus Y$, and since Y is not open, $X \setminus Y$ is not closed.

7. A topological space in which limits of sequences are not unique.

First example: Any space with the trivial topology and consisting of more than one point has this property since in this space every sequence converges to every point.

Second example: Let X be an infinite set, and let $\mathcal{O}$ consist of $\emptyset$ and the complements of finite subsets of X. Then every sequence of *distinct* points of X converges to every member of X.

8. A separable space with a nonseparable subspace.

First example: Let $(\mathfrak{R}, \mathcal{O})$ be the space of real numbers with the topology $\mathcal{O}$ generated by the base consisting of sets of the form

$$\{x\} \cup (\mathfrak{Q} \cap (x - \varepsilon, x + \varepsilon)), \qquad x \in \mathfrak{R}, \qquad \varepsilon > 0,$$

and let $(Y, \mathfrak{J})$ be the subspace of irrational numbers with the discrete topology (this *is* a subspace since every one-point set in Y is the intersection of Y and a member of Θ). Then $\mathbb{Q}$ is a countable dense subset of $(\mathfrak{R}, \Theta)$, but $(Y, \mathfrak{J})$ has no countable dense subset.

Second example: Let (X, Θ) be the space of all points (x, y) of the Euclidean plane such that $y \geqq 0$, and let Θ be the topology generated by the base consisting of sets of the following two types:

$$\{(x, y) \mid [(x - a)^2 + (y - b)^2]^{1/2} < \min (b, \varepsilon)\},$$

$$a \in \mathfrak{R}, \qquad b > 0, \qquad \varepsilon > 0,$$

$$\{(a, 0)\} \cup \{(x, y) \mid (x - a)^2 + (y - \varepsilon)^2 < \varepsilon^2\}, \qquad a \in \mathfrak{R}, \qquad \varepsilon > 0.$$

Then the set $\{(x, y) \mid x \in \mathbb{Q}, y \in \mathbb{Q} \cap \mathcal{P}\}$ is a countable dense subset of (X, Θ), but the space $\{(x, y) \mid x \in \mathfrak{R}, y = 0\}$ with the discrete topology is a subspace of (X, Θ) with no countable dense subset. (Cf. [1], p. 29, 5°.)

9. A separable space not satisfying the second axiom of countability.

Each example under Example 8 satisfies these specifications since (1) every space satisfying the second axiom of countability is separable and (2) every subspace of a space satisfying the second axiom of countability also satisfies the second axiom of countability. If either example under Example 8 satisfied the second axiom of countability, then the subspace under consideration would be separable.

10. For a given set, two distinct topologies that have identical convergent sequences.

First example: Let (X, Θ) be any uncountable space with Θ consisting of $\emptyset$ and complements of countable (possibly empty or finite) sets. Then the sequence $\{x_n\}$ converges to x iff $x_n = x$ for $n >$ some $m \in \mathfrak{N}$. In other words, the convergent sequences are precisely those of $(X, \mathfrak{J})$, where $\mathfrak{J}$ is the discrete topology. Finally, $\Theta \neq \mathfrak{J}$.

Second example: Let X be the set of all ordinal numbers less than or equal to Ω, where Ω is the first ordinal that corresponds to an uncountable set (cf. [20] and [46]). Let Θ be generated by the intervals

$$[1, \beta), \qquad (\alpha, \beta), \qquad (\alpha, \Omega],$$

where α and $\beta \in X$. Since every countable set in $X \setminus \{\Omega\}$ has an upper bound in $X \setminus \{\Omega\}$, no sequence of points in $X \setminus \{\Omega\}$ can converge to Ω. Therefore a sequence in X converges to Ω iff all but a finite number of its terms are equal to Ω. In other words, the convergent sequences in X are the same as those in the topology obtained by adjoining to the *subspace* $X \setminus \{\Omega\}$ of X the point Ω as an *isolated point* (that is, with $\{\Omega\}$ a one-point open set of the new space X).

Third example: (Cf. [4] and [29] for definitions and discussion.) Let X be the Banach space l_1 of real (alternatively, complex) sequences $x = \{x_n\}$ such that $\sum_{n=1}^{+\infty} |x_n| < +\infty$, with norm $\| x \| \equiv \sum_{n=1}^{+\infty} |x_n|$. The **strong topology** of X is that of the metric space (X, d) with $d(x, y) \equiv \| x - y \|$.

We now define a second topology for X, called the **weak topology**, in terms of the following neighborhood system:

$$N_x \equiv \left\{ y = \{y_n\} \,\middle|\, \left| \sum_{n=1}^{+\infty} a_{mn}(y_n - x_n) \right| < \varepsilon, \qquad m = 1, 2, \cdots, p \right\},$$

where (a_{mn}) is a bounded $p \times +\infty$ matrix, $x \in X$, and $\varepsilon > 0$. It can be shown that $\{N_x\}$ satisfies the conditions, specified in the Introduction, that guarantee the generation of a topological space (X, Θ).

To demonstrate that the strong topology of X is indeed stronger than the weak topology we show that every weak neighborhood of a point x contains a spherical neighborhood of x. This is an easy consequence of the triangle inequality for real series:

$$\left| \sum_{n=1}^{+\infty} a_{mn}(y_n - x_n) \right| \leqq K \cdot \sum_{n=1}^{+\infty} |y_n - x_n| = K \cdot \| y - x \|,$$

where K is an upper bound of the set of absolute values of the elements of the matrix (a_{mn}). To prove that the strong topology is *strictly* stronger than the weak topology, we shall now show that *every weak neighborhood is unbounded in the metric of the strong topology* (and hence *no* weak neighborhood is a subset of *any* strong neighborhood). Accordingly, let (a_{mn}) be the $p \times +\infty$ matrix for a weak neighborhood N_x, let $(z_1, z_2, \cdots, z_{p+1})$ be a *nontrivial* $(p + 1)$-dimensional vector such that

$$\sum_{n=1}^{p+1} a_{mn} z_n = 0 \qquad \text{for} \qquad m = 1, 2, \cdots, p,$$

and let $z_{p+2} = z_{p+3} = \cdots = 0$. The vector $y(\alpha) \equiv x + \alpha z = \{y_n(\alpha)\} = \{x_n + \alpha z_n\}$ belongs to N_x for every real number α:

$$\sum_{n=1}^{+\infty} a_{mn}(y_n(\alpha) - x_n) = \sum_{n=1}^{+\infty} a_{mn}\, \alpha z_n$$

$$= \alpha \sum_{n=1}^{p+1} a_{mn}\, z_n = 0, \qquad m = 1, 2, \cdots, p.$$

On the other hand, $\| y(\alpha) - x \| = \| \alpha z \| = | \alpha | \cdot \| z \|$, and $\| z \| \neq 0$.

We now turn our attention to sequences of points in X. We already know that every sequence $\{x^{(m)}\}$ of points in X that converges to x in the strong topology must converge to x in the weak topology. We shall now show the converse: *If* $\lim_{m \to +\infty} x^{(m)} = x$ *in the weak topology, then* $\lim_{m \to +\infty} x^{(m)} = x$ *in the strong topology.* It will then follow that the weak and strong topologies of X determine identical convergent sequences.

Assume that there exists a sequence converging to x in the weak topology but not in the strong topology. By the linear character of the two limit definitions we may assume without loss of generality that the limit x is the zero vector 0. Furthermore, if the sequence under consideration does not converge to 0 in the strong topology, then there must be a subsequence whose norms are bounded from 0. If this subsequence is denoted $\{x^{(m)}\}$, then there exists a positive number ε such that

$$\| x^{(m)} \| \geqq 5\varepsilon$$

for $m = 1, 2, \cdots$. Since $\{x^{(m)}\}$ is a subsequence of a sequence converging weakly to 0, $\{x^{(m)}\}$ must also converge weakly to 0. If we represent $x^{(m)}$:

$$x^{(m)} = (x_1^{(m)}, x_2^{(m)}, \cdots, x_n^{(m)}, \cdots),$$

then the sequence $\{x^{(m)}\}$ is represented by an infinite matrix M each row of which corresponds to one of the vectors of the sequence $\{x^{(m)}\}$. The next step is to show that since $\{x^{(m)}\}$ converges to 0 in the weak topology, every column of M is a real sequence with limit 0; that is, the limit as $m \to +\infty$ of the nth coordinate of $x^{(m)}$, for any fixed $n = 1, 2, \cdots$, is 0:

$$\lim_{m \to +\infty} x_n^{(m)} = 0.$$

This follows immediately from considering the neighborhood of 0 given by the matrix (a_{mn}) consisting of one row only, with 1 in the nth place and 0's elsewhere.

We can now define by induction two strictly increasing sequences of positive integers, $m_1 < m_2 < \cdots$ and $n_1 < n_2 < \cdots$ such that for $j \in \mathfrak{N}$:

(a)
$$\sum_{n=1}^{n_j} |x_n^{(m_j)}| < \varepsilon, \qquad \sum_{n=n_{j+1}+1}^{+\infty} |x_n^{(m_j)}| < \varepsilon,$$

and consequently

(b)
$$\sum_{n=n_j+1}^{n_{j+1}} |x_n^{(m_j)}| > 3\varepsilon.$$

Finally, we define the sequence $\{a_n\}$:

$$a_n \equiv \begin{cases} 1 & \text{if } 1 \leq n < n_1, \\ \operatorname{sgn} x_n^{(m_j)} & \text{if } n_j + 1 \leq n \leq n_{j+1}, \end{cases}$$

for $j = 1, 2, \cdots$. If N_0 is the neighborhood of the zero vector 0, defined by ε and the matrix (a_{mn}) consisting of one row only, made up of the terms of $\{a_n\}$:

$$N_0 \equiv \left\{ y = \{y_n\} \ \Big| \ \Big| \sum_{n=1}^{+\infty} a_n y_n \Big| < \varepsilon \right\},$$

then *no* point $x^{(m_j)}$ of the subsequence $\{x^{(m_j)}\}$ of $\{x^{(m)}\}$ is a member of N_0:

$$\Big| \sum_{n=1}^{+\infty} a_n x_n^{(m_j)} \Big| \geq \sum_{n=n_j+1}^{n_{j+1}} |x_n^{(m_j)}|$$
$$- \Big| \sum_{n=1}^{n_j} a_n x_n^{(m_j)} \Big| - \Big| \sum_{n=n_{j+1}+1}^{+\infty} a_n x_n^{(m_j)} \Big|$$
$$> 3\varepsilon - \varepsilon - \varepsilon = \varepsilon.$$

But this means that $\{x^{(m)}\}$ cannot converge to 0. (Contradiction.)

11. A topological space X, a set $A \subset X$, and a limit point of A that is not a limit of any sequence in A.

First example: Let X be the space of the first example under 10, preceding, with A any uncountable proper subset of X. Then any point of $X \setminus A$ is a limit point of A, but since x cannot be a limit of a sequence of points of A in the discrete topology, it cannot in the topology described in that example.

Second example: Let X be the space of the second example under 10, preceding, with $A \equiv X \setminus \{\Omega\}$ in the first topology described in that example. Then Ω is a limit point of A but no sequence in A can converge to Ω.

Third example: (Cf. [51], where a similar example is constructed. Also cf. [4], [15], [29], and [30] for definitions and discussions.) Let X be real (alternatively, complex) sequential **Hilbert space** l_2 consisting of all sequences $x = \{x_n\}$ of real numbers such that $\sum_{n=1}^{+\infty} x_n{}^2 < +\infty$ ($\sum_{n=1}^{+\infty} |x_n|^2 < +\infty$ in the complex case), and in which there is defined an **inner product** for any two points $x = \{x_n\}$ and $y = \{y_n\}$:

$$(x, y) \equiv \sum_{n=1}^{+\infty} x_n y_n \qquad \left(\sum_{n=1}^{+\infty} x_n \bar{y}_n \text{ in the complex case}\right),$$

with the properties, for x, y, and $z \in l_2$ and $\lambda \in \Re$ (only the real case will be considered henceforth in this example):

(i) $(x + y, z) = (x, z) + (y, z)$,

(ii) $(\lambda x, y) = \lambda(x, y)$,

(iii) $(y, x) = (x, y)$,

(iv) $(x, x) \geqq 0$,

(v) if $\| x \| \equiv (x, x)^{1/2}$, then $\| \; \; \|$ is a norm for which l_2 is a Banach space.

The space $X = l_2$ is now made into a topological space $(X, \mathcal{O})$ by means of a *neighborhood system*, defined somewhat as in the third example under Example 10:

$$N_x \equiv \{y \mid b \in B \Rightarrow | (y - x, b) | < \varepsilon\},$$

where B is any nonempty finite set of points of X, $x \in X$, and $\varepsilon > 0$.

Let A be the set $\{x^{(m)}\}$, where $x^{(m)}$ is the point whose mth coordinate is $\sqrt{m}$ and all other coordinates are equal to 0:

$$x^{(m)} = (0, 0, \cdots, 0, \sqrt{m}, 0, \cdots).$$

We shall show first that the zero vector 0 is a limit point of A by assuming that the neighborhood of 0,

$$N_0 \equiv \{y \mid b \in B \Rightarrow | (y, b) | < \varepsilon\},$$

where $\varepsilon > 0$ and B consists of the points $b^{(1)} = \{b_n{}^{(1)}\}$, $\cdots$, $b^{(p)} = \{b_n{}^{(p)}\}$, contains *no* point $x^{(m)}$. This means that

$$\forall\ m \in \mathfrak{N}\ \exists\ k \in \{1, 2, \cdots, p\}\ \ni\ |\sqrt{m}\overline{b}_m{}^{(k)}|\ \geqq\ \varepsilon,$$

and hence:

$$\sum_{k=1}^{p} \sum_{m=1}^{+\infty} (b_m{}^{(k)})^2 = \sum_{m=1}^{+\infty} \sum_{k=1}^{p} (b_m{}^{(k)})^2 \geqq \sum_{m=1}^{+\infty} \frac{\varepsilon^2}{m} = +\infty,$$

in contradiction to the assumed convergence of $\sum_{m=1}^{+\infty} (b_m{}^{(k)})^2$ for $k = 1, 2, \cdots, p$.

Finally, we shall show that no sequence of points of A can converge to the zero vector 0 . It is easy to show that if

$$x^{(m_1)}, x^{(m_2)}, \cdots, x^{(m_j)}, \cdots \rightarrow 0,$$

then the sequence m_1, m_2, $\cdots$ is unbounded, and we can therefore assume without loss of generality that $m_1 < m_2 < \cdots$ and

$$\sum_{j=1}^{+\infty} \frac{1}{m_j} < +\infty.$$

We conclude by defining a neighborhood N_0 of 0 in terms of $\varepsilon \equiv 1$ and the set B consisting of the single vector b whose m_jth coordinate is $1/\sqrt{m_j}$ for $j = 1, 2, \cdots$ and whose other coordinates are all 0. Then *no* point of the sequence $\{x^{(m_j)}\}$ can belong to N_0 since $(x^{(m_j)}, b) = 1$ for $j = 1, 2, \cdots$.

Fourth example: A fourth example is given below (Example 12).

Note that the phenomenon illustrated in the examples of this set cannot occur in a metric space, and therefore each of the spaces described above is neither metric nor metrizable.

12. A topological space X whose points are functions, whose topology corresponds to pointwise convergence, and which is not metrizable.

Let $(X, \mathcal{O})$ be the space of all real-valued continuous functions with domain $[0, 1]$, and let $\mathcal{O}$ be generated by the neighborhood system

$$N_f \equiv \{g \mid x \in F \Rightarrow |g(x) - f(x)| < \varepsilon\},$$

where F is a finite nonempty subset of $[0, 1]$, $f \in X$, and $\varepsilon > 0$. Clearly, if $g_n \rightarrow g$, as $n \rightarrow +\infty$, in this topology then $g_n(x) \rightarrow g(x)$, as $n \to +\infty$, for each $x \in [0, 1]$ since F can be taken to be the one-point set $\{x\}$. On the other hand, if $g_n(x) \to g(x)$, as $n \to +\infty$, for each $x \in [0, 1]$, then $g_n \to g$, as $n \to +\infty$, since for every $\varepsilon > 0$ and finite subset F of $[0, 1]$, n can be chosen sufficiently large to ensure $|\, g_n(x) - g(x) \,| < \varepsilon$ for *every* $x \in F$.

Let A be the set of all functions f in X such that:

(a) $x \in [0, 1] \Rightarrow 0 \leqq f(x) \leqq 1$,

(b) $\mu(\{x \mid f(x) = 1\}) \geqq \frac{1}{2}$.

Then 0 is a limit point of A, but if a sequence $\{f_n\}$ of members of A converged to 0 in the topology Θ, then $\{f_n(x)\}$ would converge to 0 for every $x \in [0, 1]$, and by the Lebesgue dominated convergence theorem, $\int_0^1 f_n(x)\, dx \to 0$ as $n \to +\infty$, in contradiction to the inequality $\int_0^1 f_n(x)\, dx \geqq \frac{1}{2}$. By the final remark included with Example 11, X is not metrizable.

13. A mapping of one topological space onto another that is continuous but neither open nor closed.

First example: Let $f(x) \equiv e^x \cos x$, with domain and range $\mathfrak{R}$, with the standard topology. Then f is continuous, but $f((-\infty, 0))$ is not open and $f(\{-n\pi \mid n \in \mathfrak{N}\})$ is not closed.

Second example: Let X be the space $\mathfrak{R}$ with the discrete topology, let Y be the space $\mathfrak{R}$ with the standard topology, and let the mapping be the identity mapping.

14. A mapping of one topological space onto another that is open and closed but not continuous.

First example: Let X be the unit circle $\{(x, y) \mid x^2 + y^2 = 1\}$ with the topology inherited from the standard topology of the plane, let Y be the half-open interval $[0, 2\pi)$ with the topology inherited from the standard topology of $\mathfrak{R}$, and let the mapping f be $(x, y) \to \theta$, where $x = \cos\theta$, $y = \sin\theta$, and $0 \leqq \theta < 2\pi$. Then f is both open and closed since its inverse is continuous, but f is discontinuous at $(1, 0)$.

Second example: The inverse of the mapping of the second example under Example 13.

15. A mapping of one topological space onto another that is closed but neither continuous nor open.

Let X be the unit circle $\{(x, y) \mid x^2 + y^2 = 1\}$ with the topology inherited from the standard topology of the plane, let Y be the half-open interval $[0, \pi)$ with the topology inherited from the standard topology of $\Re$, and let the mapping f be defined:

$$(\cos \theta, \sin \theta) \to \begin{cases} 0 & \text{if} \quad 0 \leq \theta \leq \pi, \\ \theta - \pi & \text{if} \quad \pi < \theta < 2\pi. \end{cases}$$

Then f is not open since the open upper semicircle of X maps onto a point, and f is not continuous at $(1, 0)$ (cf. the first example under Example 14). However f is closed, as we shall now see. Assume that f is *not* closed. Then there is a closed set A of X such that $B \equiv f(A)$ is not closed in Y. Therefore there is a sequence $\{b_n\}$ of points of B such that $b_n \to b$, and $b \notin B$. If $f(p_n) = b_n$, where $p_n \in A$, for $n = 1, 2, \cdots$, since A is compact we may assume without loss of generality that $\{p_n\}$ converges: $p_n \to p \in A$. Since $f(p_n) \to b \neq f(p)$, f is discontinuous at p, and $p = (1, 0)$. But this means that there exists a subsequence of $\{p_n\}$ approaching $(1, 0)$ from either the upper or the lower semicircle; in the former case $b_n \to 0 \in B$, and in the latter case $\{b_n\}$ cannot converge in Y. In either case a contradiction is obtained, and f is therefore closed.

16. A mapping of one topological space onto another that is continuous and open but not closed.

Let (X, Θ) be the Euclidean plane with the standard topology, let $(Y, \mathfrak{I})$ be $\Re$ with the usual topology, and let the mapping be the projection $P: (x, y) \to x$. Then P is clearly continuous and open, but $P(\{(x, y) \mid y = 1/x > 0\})$ is not closed in $(Y, \mathfrak{I})$.

17. A mapping of one topological space onto another that is open but neither continuous nor closed.

First example: Let $X = Y = \Re$ with the standard topology and let f be the function defined in Example 27, Chapter 8, whose range on every nonempty open interval is $\Re$. This function is clearly open since the image of every nonempty open set is $\Re$, and it is everywhere discontinuous. To show that f is not closed, let x_n be a point between n and $n + 1$ such that $f(x_n)$ is between $1/(n + 1)$ and $1/n$, for $n = 1, 2, \cdots$. Then $\{x_n\}$ is closed and $\{f(x_n)\}$ is not.

Second example: Let (X, Θ) be the plane with Θ consisting of $\emptyset$ and complements of countable sets, let $(Y, \mathfrak{I})$ be $\Re$ with $\mathfrak{I}$ consisting of $\emptyset$ and complements of finite sets, and let the mapping be the projection $P : (x, y) \to x$. Then P is open since any nonempty open set in (X, Θ) must contain some horizontal line, whose image is $\Re$. On the other hand, P is not closed since the set of points $(n, 0)$, where $n \in \Re$, is closed in (X, Θ), but its image is not closed in $(Y, \Im)$, and since the inverse image of any open set in $(Y, \Im)$ that is a proper subset of Y cannot be open in (X, Θ), P is not continuous.

18. A mapping of one topological space onto another that is continuous and closed but not open.

Let X and Y be the closed interval $[0, 2]$ with the usual topology, and let

$$f(x) \equiv \begin{cases} 0 & \text{if } 0 \leq x \leq 1, \\ x - 1 & \text{if } 1 < x \leq 2. \end{cases}$$

Then f is clearly continuous, and hence closed since X and Y are compact metric spaces. On the other hand, $f((0, 1))$ is not open in Y.

19. A topological space X, and a subspace Y in which there are two disjoint open sets not obtainable as intersections of Y with disjoint open sets of X.

Let $X \equiv \Re$, the open sets being $\emptyset$ or complements of finite sets, and let $Y \equiv \{1, 2\}$. Then $\{1\} = Y \cap (X \setminus \{2\})$ and $\{2\} = Y \cap (X \setminus \{1\})$, so that the subspace topology of Y is discrete. However, the two disjoint open sets $\{1\}$ and $\{2\}$ of Y are not the intersections of Y with *disjoint* open sets of X since no two nonempty open sets of X are disjoint.

20. Two nonhomeomorphic topological spaces each of which is a continuous one-to-one image of the other.

First example: Let X and Y be subspaces of $\Re$, where $\Re$ has the standard topology, defined:

$$X \equiv \bigcup_{n=0}^{+\infty} ((3n, 3n+1) \cup \{3n+2\}), \qquad Y \equiv (X \setminus \{2\}) \cup \{1\}.$$

Let the mappings S of X onto Y and T of Y onto X be defined

$$S(x) \equiv \begin{cases} x & \text{if } x \neq 2, \\ 1 & \text{if } x = 2, \end{cases} \qquad T(y) \equiv \begin{cases} \frac{1}{2}y & \text{if } y \leq 1, \\ \frac{1}{2}y - 1 & \text{if } 3 < y < 4, \\ y - 3 & \text{if } y \geq 5. \end{cases}$$

Then S and T are continuous, one-to-one, and onto mappings. However, X and Y are not homeomorphic since under any homeomorphism of Y onto X the point 1 of Y can have no correspondent.

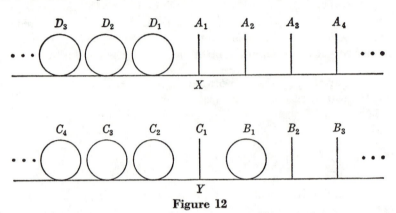

Figure 12

Second example: Let X and Y be subsets of the plane, with the standard topology, as indicated in Figure 12. The vertical segments are of length 2 and are open at the top ends, and the circles are of radius 1. The mapping S of X onto Y is defined as follows: The horizontal line of X is mapped onto the horizontal line of Y, the cirles D_n onto the circles C_{n+1}, and the segments A_{n+2} onto the segments B_{n+1}, $n \in \mathfrak{N}$, by a translation downward in Figure 12, and the segment A_2 is mapped onto the circle B_1 by a formula of the type $x = \sin \pi t$, $y = 1 - \cos \pi t$, where $0 \leq t < 2$. The segment A_1 is mapped by downward translation onto C_1. The mapping T of Y onto X is defined as follows: The horizontal line of Y is mapped onto the horizontal line of X, the circles C_{n+1} onto the circles D_{n+2}, the segments B_{n+1} onto the segments A_n, $n \in \mathfrak{N}$, and the circle B_1 onto the circle D_1 by a translation upward and to the left in Figure 12, and the segment C_1 is mapped onto the circle D_2 by a formula of the type used above. The mappings S and T have the properties claimed. It is left as an exercise for the reader to show that the spaces X and Y are not homeomorphic.

21. A decomposition of a three-dimensional Euclidean ball B into five disjoint subsets S_1, S_2, S_3, S_4, S_5 (where S_5 consists

of a single point) and five rigid motions, R_1, R_2, R_3, R_4, R_5 such that

$$B \cong R_1(S_1) \cup R_2(S_2) \cong R_3(S_3) \cup R_4(S_4) \cup R_5(S_5)$$

(where "$\cong$" means "is congruent to").
(Cf. references given below.)

22. For ε, $M > 0$, two Euclidean balls B_ε and B_M of radius ε and M respectively, a decomposition of B_ε into a finite number of disjoint subsets S_1, S_2, $\cdots$, S_n, and n rigid motions R_1, R_2, $\cdots$, R_n such that

$$B_M = R_1(S_1) \cup R_2(S_2) \cup \cdots \cup R_n(S_n).$$

The last two examples are due to the work of Hausdorff, Banach, Tarski, von Neumann, and R. Robinson. We shall give only a reference where these are discussed in detail. (Cf. [41].)

Chapter 13
Function Spaces

Introduction

A **function space** is a collection of functions having a common domain D. It is usually assumed that a function space is endowed with some sort of algebraic or topological structure. In this chapter we shall focus our attention on the *algebraic* structures of certain spaces of real-valued functions of a real variable whose common domain is a fixed interval I.

A function space S of real-valued functions on an interval I is said to be a **vector space** or a **linear space over** $\Re$ (the real-number system) iff S is closed with respect to linear combinations with real coefficients; that is, iff

$$f, g \in S, \lambda, \mu \in \Re \Rightarrow \lambda f + \mu g \in S,$$

where the function λf is defined

$$(\lambda f)(x) \equiv \lambda(f(x)).$$

It is easy to show that a function space of real-valued functions on an interval is a vector space iff it is closed with respect to the two operations of *addition*, $f + g$, and *scalar multiplication*, λf. The abstract concept of *vector space* is defined axiomatically in the Introduction to Chapter 12. (For further discussion cf. [22].) Many of the most important classes of functions in analysis are linear spaces over $\Re$ (or over $\mathfrak{C}$, the field of complex numbers, in which case the coefficients λ and μ are arbitrary complex numbers). Examples of spaces of real-valued functions that are linear spaces over $\Re$ are:

1. *All* (real-valued) functions on an interval I. 2. All *bounded* functions on an interval I. 3. All *Riemann-integrable* functions on a closed interval $[a, b]$. 4. All *Lebesgue-measurable* functions on an interval I. 5. All *Lebesgue-integrable* functions on an interval I. 6. All Lebesgue-measurable functions on an interval I the pth power of whose absolute value is Lebesgue-integrable on I, where $p \geq 1$. 7. All *continuous* functions on an interval I. 8. All *sectionally continuous* functions on a closed interval $[a, b]$ (cf. [34], p. 131). 9. All *sectionally smooth* functions on a closed interval $[a, b]$ (cf. [34], p. 131). 10. All functions having kth order derivatives at every point of an interval I for every k not exceeding some fixed positive integer n. 11. All functions having a *continuous* kth order derivative on an interval I for every k not exceeding some fixed positive integer n. 12. All *infinitely differentiable* functions on an interval I. 13. All (*algebraic*) *polynomials* on an interval I. 14. All (algebraic) polynomials, on an interval I, of degree not exceeding some fixed positive integer n. 15. All *trigonometric polynomials*, on an interval I, having the form

(1)
$$\sum_{i=0}^{n} \alpha_i \cos ix + \beta_i \sin ix,$$

where n is arbitrary. 16. All trigonometric polynomials of the form (1) where n is fixed. 17. All *step-functions* on a closed interval $[a, b]$. 18. All *constant functions* on an interval I. 19. All functions satisfying a given linear homogeneous differential equation, such as

$$\frac{d^3 y}{dx^3} + \sin x \frac{d^2 y}{dx^2} + e^x \frac{dy}{dx} - xy = 0,$$

on an interval I. Nineteen more examples of linear spaces over $\mathfrak{R}$ are provided by permitting the functions of the preceding spaces to be complex-valued functions.

In spite of the vital role played by linearity in analysis, there are several important classes of real-valued functions that do *not* form linear spaces. Some of these are indicated in the first five examples below, which can be interpreted as saying that the following spaces of real-valued functions on a fixed interval are nonlinear: (*i*) all monotonic functions on $[a, b]$, (*ii*) all periodic functions on $(-\infty, +\infty)$ (*iii*) all semicontinuous functions on $[a, b]$, (*iv*) all functions whose squares are Riemann-integrable on $[a, b]$, (*v*) all functions whose squares are Lebesgue-integrable on $[a, b]$.

II. Higher Dimensions

A function space S of real-valued functions on an interval I is called an **algebra** over $\mathfrak{R}$ iff it is closed with respect both to linear combinations with real coefficients and to products; that is, iff S is a linear space over $\mathfrak{R}$ and

$$f \in S, g \in S \Rightarrow fg \in S.$$

(As with linear spaces, the abstract concept of *algebra* is defined by means of axioms. Cf. [22], vol. 2, pp. 36, 225.) As a consequence of the identity

$$(2) \qquad fg = \tfrac{1}{4}(f + g)^2 - \tfrac{1}{4}(f - g)^2$$

it follows that a function space that is a linear space is an algebra iff it is closed with respect to squaring.

A function space S of real-valued functions on an interval I is called a **lattice** iff it is closed with respect to the formation of the two binary operations of *join* and *meet*, defined and denoted:

join of f and g : $\quad f \vee g, \qquad (f \vee g)(x) \equiv \max\,(f(x)\,,\, g(x)),$

meet of f and g : $\quad f \wedge g, \qquad (f \wedge g)(x) \equiv \min\,(f(x)\,,\, g(x)).$

(Again, the abstract concept of lattice is defined axiomatically. Cf. [9].) For a given real-valued function f, the two nonnegative functions f^+ and f^- are defined and related to f and its absolute value $|f|$ as follows:

$$(3) \qquad f^+ \equiv f \vee 0, \qquad\qquad f^- \equiv (-f) \vee 0,$$

$$(4) \qquad f = f^+ - f^-, \qquad\qquad |f| = f^+ + f^-,$$

$$(5) \qquad f^+ = \tfrac{1}{2}|f| + \tfrac{1}{2}f, \qquad\qquad f^- = \tfrac{1}{2}|f| - \tfrac{1}{2}f.$$

Thanks to these relationships and the additional ones that follow:

$$(6) \qquad f \vee g = -[(-f) \wedge (-g)],$$

$$(7) \qquad f \wedge g = -[(-f) \vee (-g)],$$

$$(8) \qquad f \vee g = \tfrac{1}{2}(f + g) + \tfrac{1}{2}|f - g|,$$

$$(9) \qquad f \wedge g = \tfrac{1}{2}(f + g) - \tfrac{1}{2}|f - g|,$$

a function space that is a linear space is a lattice iff it is closed with respect to *any one* of the following five binary or unary operations:

(10) $$f \vee g, f \wedge g,$$

(11) $$f^+, f^-, |f|.$$

In the preceding list of linear spaces, those that are also both algebras and lattices are 1, 2, 3, 4, 7, 8, 17, and 18. Those that are neither algebras nor lattices are 14 (cf. Example 6, below), 16, and 19. Those that are algebras and not lattices are 9, 10, 11 (cf. Example 7, below), 12, 13, and 15. Those that are lattices and not algebras are 5 (cf. Example 8, below) and 6.

1. Two monotonic functions whose sum is not monotonic.

$$\sin x + 2x \text{ and } \sin x - 2x \quad \text{on } [-\pi, \pi].$$

2. Two periodic functions whose sum is not periodic.

$$\sin x \text{ and } \sin \alpha x, \quad \alpha \text{ irrational, on } (-\infty, +\infty).$$

If $\sin x + \sin \alpha x$ were periodic with nonzero period p, then the following identities would hold for all real x:

$$\sin (x + p) + \sin (\alpha x + \alpha p) = \sin x + \sin \alpha x,$$

$$\sin (x + p) - \sin x = -[\sin (\alpha x + \alpha p) - \sin \alpha x],$$

$$\cos (x + \tfrac{1}{2}p) \sin (\tfrac{1}{2}p) = -\cos (\alpha x + \tfrac{1}{2}\alpha p) \sin (\tfrac{1}{2}\alpha p),$$

$$\cos x \sin (\tfrac{1}{2}p) = -\cos (\alpha x) \sin (\tfrac{1}{2}\alpha p).$$

If x is set equal to $\tfrac{1}{2}\pi$, the left-hand side of this last equation vanishes, and hence $\sin \tfrac{1}{2}\alpha p = 0$, and αp is a multiple of 2π. If αx is set equal to $\tfrac{1}{2}\pi$, the right-hand side vanishes and hence $\sin \tfrac{1}{2}p = 0$, and p is a multiple of 2π. Since α is irrational this is impossible and the desired contradiction has been reached. (Cf. [36], p. 550, Note.)

3. Two semicontinuous functions whose sum is not semicontinuous.

If

$$f(x) \equiv \begin{cases} 1 & \text{if } x > 0, \\ 2 & \text{if } x = 0, \\ -1 & \text{if } x < 0, \end{cases} \qquad g(x) \equiv \begin{cases} 1 & \text{if } x > 0, \\ -2 & \text{if } x = 0, \\ -1 & \text{if } x < 0, \end{cases}$$

then $f(x) + g(x)$ fails to be semicontinuous at $x = 0$, although $f(x)$ is everywhere upper semicontinuous and $g(x)$ is everywhere lower semicontinuous.

More dramatic examples are possible if f and g are functions each of which is semicontinuous everywhere but *upper* semicontinuous at *some* points and *lower* semicontinuous at *other* points. In the following examples the notation p/q will indicate a quotient of integers in lowest terms, with $q > 0$, the number 0 being represented $0/1$. If

$$f(x) \equiv \begin{cases} 1 & \text{if } x = p/q, q \text{ odd,} \\ 0 & \text{otherwise,} \end{cases} \qquad g(x) \equiv \begin{cases} -1 & \text{if } x = p/q, q \text{ even,} \\ 0 & \text{otherwise,} \end{cases}$$

then

$$f(x) + g(x) = \begin{cases} 1 & \text{if } x = p/q, q \text{ odd,} \\ -1 & \text{if } x = p/q, q \text{ even,} \\ 0 & \text{if } x \text{ is irrational,} \end{cases}$$

and the function $f(x) + g(x)$ is semicontinuous iff x is rational, and hence *almost nowhere* (that is, $f + g$ *fails* almost everywhere to be semicontinuous).

Now consider the three functions defined as follows:

$$F(x) \equiv \begin{cases} 4/q & \text{if } x = p/q, \ q \text{ odd,} \\ -2 - (4/q) & \text{if } x = p/q, \ q \text{ even,} \\ -2 & \text{if } x \text{ is irrational,} \end{cases}$$

$$G(x) \equiv \begin{cases} -1 - (1/q) & \text{if } x = p/q, \ q \text{ odd,} \\ 1 + (1/q) & \text{if } x = p/q, \ q \text{ even,} \\ -1 & \text{if } x \text{ is irrational.} \end{cases}$$

$$H(x) \equiv \begin{cases} -1 - (1/q) & \text{if } x = p/q, \ q \text{ odd,} \\ 3 + (1/q) & \text{if } x = p/q, \ q \text{ even,} \\ 3 & \text{if } x \text{ is irrational.} \end{cases}$$

Then F, G, and H are individually semicontinuous everywhere, while their sum,

$$F(x) + G(x) + H(x) = \begin{cases} -2 + (2/q) & \text{if } x = p/q, \ q \text{ odd,} \\ 2 - (2/q) & \text{if } x = p/q, \ q \text{ even,} \\ 0 & \text{if } x \text{ is irrational,} \end{cases}$$

is *nowhere* semicontinuous.

4. Two functions whose squares are Riemann-integrable and the square of whose sum is not Riemann-integrable.

If

$$f(x) \equiv \begin{cases} 1 & \text{if } x \text{ is irrational,} \\ -1 & \text{if } x \text{ is rational,} \end{cases}$$

and

$$g(x) \equiv \begin{cases} 1 & \text{if } x \text{ is algebraic,} \\ -1 & \text{if } x \text{ is transcendental,} \end{cases}$$

then

$$f(x) + g(x) = \begin{cases} 2 & \text{if } x \text{ is algebraic and irrational,} \\ 0 & \text{otherwise.} \end{cases}$$

Then f^2 and g^2 are constants and hence integrable on every closed interval — e.g., on [0, 1] to be specific — while $(f + g)^2$ is everywhere discontinuous and thus Riemann-integrable on *no* interval, and in particular it is not integrable on [0, 1].

5. Two functions whose squares are Lebesgue-integrable and the square of whose sum is not Lebesgue-integrable.

Let E_1 be a nonmeasurable subset of [0, 1] and let E_2 be a non-measurable subset of [2, 3] (cf. Example 10, Chapter 8). Then let

$$f(x) \equiv \begin{cases} 1 & \text{if } x \in [0, 1] \cup E_2, \\ -1 & \text{if } x \in [2, 3] \setminus E_2, \\ 0 & \text{otherwise,} \end{cases}$$

and

$$g(x) \equiv \begin{cases} 1 & \text{if } x \in [2, 3] \cup E_1, \\ -1 & \text{if } x \in [0, 1] \setminus E_1, \\ 0 & \text{otherwise.} \end{cases}$$

Then

$$f(x) + g(x) = \begin{cases} 2 & \text{if } x \in E_1 \cup E_2, \\ 0 & \text{otherwise.} \end{cases}$$

Thus since

$$f^2(x) = g^2(x) = \begin{cases} 1 & \text{if } x \in [0, 1] \cup [2, 3], \\ 0 & \text{otherwise,} \end{cases}$$

and

$$(f(x) + g(x))^2 = \begin{cases} 4 & \text{if } x \in E_1 \cup E_2, \\ 0 & \text{otherwise,} \end{cases}$$

the result follows, since $E_1 \cup E_2$ is nonmeasurable.

6. A function space that is a linear space but neither an algebra nor a lattice.

The polynomials $cx + d$ of degree at most 1 on the closed interval [0, 1] form a linear space. They do not form an algebra since the square of the member x is not a member. They do not form a lattice, since although $2x - 1$ is a member, $|2x - 1|$ is not.

7. A linear function space that is an algebra but not a lattice.

The set of all functions that are continuously differentiable on [0, 1] form an algebra because of the formula $(fg)' = fg' + f'g$. However, they do not form a lattice. The function

$$f(x) \equiv \begin{cases} 0 & \text{if } x = 0, \\ x^3 \sin 1/x & \text{if } 0 < x \leqq 1 \end{cases}$$

is continuously differentiable on [0, 1], but its absolute value fails to be differentiable at the infinitely many points where $f(x) = 0$. In fact, $|f(x)|$ is not even sectionally smooth.

8. A linear function space that is a lattice but not an algebra.

The set of all functions that are Lebesgue-integrable on [0, 1] is a linear space and a lattice. However, this space is not an algebra since the function

$$f(x) \equiv \begin{cases} 0 & \text{if } x = 0, \\ x^{-1/2} & \text{if } 0 < x \leqq 1, \end{cases}$$

is a member of the set but its square is not.

9. Two metrics for the space $C([0, 1])$ of functions continuous on [0, 1] such that the complement of the unit ball in one is dense in the unit ball of the other.

Let ρ and σ be two metrics defined as follows: For $f, g \in C([0, 1])$, let

$$\rho(f, g) = \sqrt{\int_0^1 |f(x) - g(x)|^2 \, dx} \equiv \| f - g \|_2;$$

$$\sigma(f, g) = \sup_{0 \leq x \leq 1} |f(x) - g(x)| \equiv \| f - g \|_\infty.$$

Let $P \equiv \{f \mid \rho(f, 0) \leq 1\}$, $\Sigma \equiv \{f \mid \sigma(f, 0) \leq 1\}$ be the unit balls in these metrics. Clearly $\Sigma \subset P$. We shall show that the complement of Σ is dense in P. Indeed, let $f \in P$, $0 < \varepsilon < 1$. If $\| f \|_\infty > 1$, then $f \notin \Sigma$ and we need look no further. If $\| f \|_\infty \leq 1$, let $g(x)$ be defined by:

$$g(x) = \begin{cases} 0 & \text{if } 0 \leq x \leq \tfrac{1}{2} - (\varepsilon^2/9) \text{ or if } \tfrac{1}{2} + (\varepsilon^2/9) \leq x \leq 1, \\ 3 & \text{if } x = \tfrac{1}{2}, \\ \text{linear otherwise.} \end{cases}$$

Then $f(x) + g(x) \notin \Sigma$ and $\| f - (f + g) \|_2 = \| g \|_2 < \sqrt{9 \cdot (\varepsilon^2/9)} = \varepsilon$.

This last example illustrates an essential distinction between finite-dimensional and infinite-dimensional normed linear spaces. In either case the closed unit ball is such that any line through the origin (that is, all scalar multiples of a fixed nonzero point) meets the unit ball in a closed segment having the origin as midpoint. In the finite-dimensional case the topology is thereby uniquely determined. The present example shows that in the infinite-dimensional case this is not true.

Bibliography

1. Alexandrov, P., and H. Hopf, *Topologie*, Springer, Berlin (1935).
2. *American Mathematical Monthly*, **68**, 1 (January, 1961), p. 28, Problem 3.
3. ———, **70**, 6 (June-July, 1963), p. 674.
4. Banach, S., *Théorie des opérations linéaires*, Warsaw (1932).
5. Besicovitch, A. S., "Sur deux questions de l'intégrabilité," *Journal de la Société des Math. et de Phys. à l'Univ. à Perm*, **II** (1920).
6. ———, "On Kakeya's problem and a similar one," *Math. Zeitschrift*, **27** (1928), pp. 312–320.
7. ———, "On the definition and value of the area of a surface," *Quarterly Journal of Mathematics*, **16** (1945), pp. 86–102.
8. ———, "The Kakeya problem," Mathematical Association of America Film, Modern Learning Aids, New York; *American Mathematical Monthly*, **70, 7** (August-September, 1963), pp. 697–706.
9. Birkhoff, G., *Lattice theory*, American Mathematical Society Colloquium Publications, 25a (1948).
10. Boas, R. P., *A primer of real functions*, Carus Mathematical Monographs, No. 13, John Wiley and Sons, Inc., New York (1960).
11. Bourbaki, N., *Eléments de mathématique*, Première partie.
        I. "Théorie des ensembles," *Actualités scientifiques et industrielles*, 846, Paris (1939).
        II. "Topologie générale," Chapters I, II, *ibid.*, 858, Paris (1940).
        III. "Topologie générale," Chapter III, *ibid.*, 916–1143, Paris (1951).
13. Brouwer, L. E. J., "Zur Analysis Situs," *Math. Annalen*, **68** (1909), pp. 422–434.
14. Carathéodory, C., *Vorlesungen über reelle Funktionen*, 2d ed., G. B. Teubner, Leipzig (1927).
15. Dunford, N., and J. Schwartz, *Linear operators*, vol. 1, Interscience Publishers, New York (1958).
16. Graves, L. M., *Theory of functions of real variables*, 2d ed., McGraw-Hill Book Company, Inc., New York (1956).
17. Hall, D., and G. Spencer, *Elementary topology*, John Wiley and Sons, Inc., New York (1955).
18. Halmos, P. R., *Measure theory*, D. Van Nostrand Company, Inc., Princeton, N. J. (1950).
19. Hancock, H., *Foundations of the theory of algebraic numbers*, Macmillan, New York (1931).

20. Hausdorff, F., *Mengenlehre*, Dover Publications, New York (1944).
21. Hobson, E. W., *The theory of functions*, Harren Press, Washington (1950).
22. Jacobson, N., *Lectures in abstract algebra*, vols. 1 and 2, D. Van Nostrand Company, Inc., New York (1951).
23. Kakeya, S., "Some problems on maximum and minimum regarding ovals," *Tôhoku Science Reports*, 6 (July, 1917), pp. 71–88.
24. Kelley, J. L., *General topology*, D. Van Nostrand Company, Inc., Princeton, N. J. (1955).
25. Knaster, B., and C. Kuratowski, "Sur les ensembles connexes," *Fundamenta Mathematica*, 2 (1921), pp. 201–255.
26. Kolmogorov, A. N., "On the representation of continuous functions of several variables by superposition of continuous functions of a smaller number of variables," *Doklady Akademia Nauk SSSR* (N.S.), 108 (1956), pp. 179–182.
27. Kuratowski, C., *Topologie 1*, Warsaw (1933).
28. Lebesgue, H., *Leçons sur l'intégration*, Gauthier–Villars, Paris (1904).
29. Loomis, L., *An introduction to abstract harmonic analysis*, D. Van Nostrand Company, Inc., Princeton, N. J. (1953).
30. McShane, E. J., and T. A. Botts, *Real analysis*, D. Van Nostrand Company, Inc., Princeton, N. J. (1959).
31. Mazurkiewicz, S., *Comptes Rendus Soc. Sc. et Lettres de Varsovie*, vol. 7 (1914), pp. 322–383, especially pp. 382–383.
32. Munroe, M. E., *Introduction to measure and integration*, Addison-Wesley Publishing Company, Inc., Reading, Mass. (1953).
33. Newman, M. H. A., *Elements of the topology of plane sets of points*, Cambridge University Press, Cambridge (1939).
34. Olmsted, J. M. H., *Advanced calculus*, Appleton-Century-Crofts, Inc., New York (1961).
35. ———, *The real number system*, Appleton-Century-Crofts, Inc., New York (1962).
36. ———, *Real variables*, Appleton-Century-Crofts, Inc., New York (1956).
37. Osgood, W. F., "A Jordan curve of positive area," *Trans. Amer. Math. Soc.*, 4 (1903), pp. 107–112.
38. Pastor, J. R., *Elementos de la teoria de funciones*, 3d ed., Ibero-Americana, Madrid — Buenos Aires (1953).
39. Pollard, H., *The theory of algebraic numbers*, Carus Mathematical Monographs, No. 9, John Wiley and Sons, Inc., New York (1950).
40. Radó, T., *Length and area*, American Mathematical Society Colloquium Publications, vol. XXX, New York (1958).

II. Higher Dimensions

41. Robinson, R. M., "On the decomposition of spheres," *Fundamenta Mathematica*, **34** (1947), pp. 246–266.
42. Rudin, W., *Principles of mathematical analysis*, McGraw-Hill Book Company, Inc., New York (1953).
43. Sierpinski, W., *Bull. Internat. Ac. Sciences Cracovie* (1911), p. 149.
44. ———, "Sur un problème concernant les ensembles mesurables superficiellement, *Fundamenta Mathematica*, 1 (1920), pp. 112–115.
45. ———, *General topology*, University of Toronto Press, Toronto (1952).
46. ———, *Cardinal and ordinal numbers*, Warsaw (1958).
47. Steinitz, E., "Bedingte konvergente Reihen und konvexe Systeme," *Journal für Mathematik*, **143** (1913), pp. 128–175.
48. Titchmarsh, E. C., *The theory of functions*, Oxford University Press, Oxford (1932).
49. Toeplitz, O., "Über allgemeine lineare Mittelbildungen," *Prace Matematyczne-Fizyczne*, **22** (1911), pp. 113–119.
50. Vaidyanathaswamy, R., *Treatise on set topology*, Part 1, Indian Mathematical Society, Madras (1947).
51. Von Neumann, J., "Zur Algebra der Funktionaloperationen und Theorie der normalen Operatoren," *Math. Annalen*, **102** (1930), pp. 370–427.
52. Zygmund, A., *Trigonometrical series*, Warsaw (1935).
53. ———, *Trigonometric series*, vol. I, Cambridge University Press, Cambridge (1959).

Special Symbols

Symbol	Meaning	Page
$\in$ ($\notin$)	is (is not) an element of	3
$\subset, \supset$	is contained in, contains	3
$\Rightarrow$	implies, only if	3
iff	if and only if	3
$\Leftrightarrow$	if and only if	3
$\{a, b, c, \cdots\}$	the set consisting of the members $a, b, c, \cdots$	3
$\{\cdots \mid \cdots\}$	the set of all $\cdots$ such that $\cdots$	3
$\equiv$	equal by definition	3
$A \cup B$	the union of A and B	3
$A \cap B$	the intersection of A and B	3
$A \setminus B$	the set of points in A and not in B	3
A'	the complement of A	4
$\varnothing$	the empty set	4
(a, b)	ordered pair	4
$A \times B$	Cartesian product	4
$\exists$	there exist(s)	4
$\ni$	such that	4
D_f, R_f	domain of f, range of f	4
f^{-1}	the function inverse to f	4
$f : A \to B, A \xrightarrow{f} B$	a function $f \ni D_f = A, R_f \subset B$	5
$f(S)$	$\{y \mid \exists x \in S \ni f(x) = y\}$	6
$f \circ g$	composite of f and g $((f \circ g)(x) = f(g(x)))$	6

Symbol	Meaning	Page
$\mathfrak{F}$	field	6
$\mathfrak{G}$	group	7
$\mathfrak{P}$	positive part of a field	7
$<, \leqq, >, \geqq$	order symbols	8
$\lvert x \rvert$	absolute value of x	9
$(a, b), [a, b)$ $(a, b], [a, b]$ $(a, +\infty), (-\infty, a)$ $[a, +\infty), (-\infty, a]$ $(-\infty, +\infty)$	intervals of an ordered system	9
$N(a, \varepsilon), D(a, \varepsilon)$	neighborhood, deleted neighborhood	9
$\max (x, y)$	maximum of x and y	9
$\min (x, y)$	minimum of x and y	9
$\inf (A), \inf A$;	infimum of A	10
$\sup (A), \sup A$	supremum of A	10
$x \leftrightarrow x'$	one-to-one correspondence	10
$\mathfrak{R}$	the real number system	10
$\operatorname{sgn} x$	signum function	10
χ_A	characteristic function of the set A	10
$\mathfrak{N}$	the natural number system	10
$\mathfrak{Q}$	the field of rational numbers	11
$\mathfrak{B}$	ring	11
$\mathfrak{D}$	integral domain	11
$\mathfrak{I}$	the ring of integers	11
$\forall$	for all	12
$\lim\limits_{x \to a} f(x)$	the limit of $f(x)$ at $x = a$	12
$f'(a)$	the derivative of $f(x)$ at $x = a$	12
$\{a_n\}$	a sequence $a_1, a_2, \cdots$	13
(x, y)	a complex number	13
$\mathfrak{C}$	the field of complex numbers	13
$x_2 + iy$	a complex number	13
$\mathfrak{K}$	the field of rational functions	16
Φ	$\{a + b\sqrt{5} \mid a, b \in \mathfrak{I}\}$	17
$m \mid n$	m divides n	18
$\bigcup\limits_{n=1}^{+\infty} A_n, \bigcap\limits_{n=1}^{+\infty} A_n$	union, intersection, of the (countably) infinite set of sets $A_1, A_2, \cdots$	20

Symbol	Meaning	Page
$F(A)$, $I(A)$, $\bar{A}$	frontier, interior, closure of A	20
$\{U_\alpha\}$	open covering	21
$[x]$	sup $\{n \mid n \in \mathfrak{N}, n \leq x\}$	21
$\overline{\lim}_{x \to a}$, $\underline{\lim}_{x \to a} f(x)$	limit superior, inferior, of $f(x)$ at $x = a$	21
$D(\pm \infty, N)$	deleted neighborhoods of $\pm \infty$	22
$\lim_{x \to \pm \infty} f(x)$	limits at $\pm \infty$	22
$\lim_{x \to a} f(x) = \pm \infty$	infinite limits	22
F_σ	F-sigma	30
$\overline{\lim}_{n \to +\infty}$, $\underline{\lim}_{n \to +\infty} A_n$	limit superior, inferior, of the sequence of sets $A_1, A_2, \cdots$	51
$\sum a_n = \sum_{n=1}^{+\infty} a_n$	infinite series	53
(t_{ij})	matrix	64
$f^{(n)}(x)$	nth derivative of $f(x)$	70
$\deg P$	degree of polynomial P	74
σ-ring	sigma-ring	83
A	class of sets	83
C	class of compact sets	83
$x + A$	translate of A by x	83
B	class of Borel (-measurable) sets	83
S	σ-ring	83
$\rho \ll \sigma$	ρ absolutely continuous with respect to σ	84
X, (X, S)	measure space	84
$\tilde{B}$	class of Lebesgue-measurable sets	84
μ	measure	84
μ_*, μ^*	inner, outer measure	84
C	the Cantor set	85
$\mathfrak{c}$	cardinality of $\mathfrak{R}$	86
$D(A)$	difference set (of A)	87
G_δ, $F_{\sigma\delta}$, $\cdots$	G-delta, F-sigma delta, $\cdots$	91
$r + A(\text{mod } 1]$	translation (of A by r) modulo 1	92
$\mathfrak{J}$	$\{z \mid z \in \mathfrak{C}, \mid z \mid = 1\}$	93
$\mathfrak{J}_0$	$\{z \mid z = e^{2\pi i \theta}, \theta \in \mathfrak{Q}, 0 \leq \theta < 1\}$	93
$\mathfrak{S}$, $\tilde{\mathfrak{S}}$	$\mathfrak{J}/\mathfrak{J}_0$, one-to-one preimage of $\mathfrak{S}$ in $\mathfrak{J}$	93
$\tilde{\mu}$	measure on $\mathfrak{J}$	94

Special Symbols

Index

(The numbers refer to pages.)

Errata

Example 30 on page 105 should be elaborated as follows.

30. A continuous strictly monotonic funtion with a vanishing derivative almost everywhere

A function with these properties is given by A. C. Zaanen and W. A. Luxemburg [3]. In mildly modified form it appeared in the original edition of this book. In the Russian translation of the original there is offered a footnote to the effect that the example cited is in error. To put the matter (hopefully) to rest, the following discussion attempts to validate the original construction.

At root is the Cantor function ϕ of Example 15. For any closed interval $[a, b]$, *its* Cantor function, denoted ψ for simplicity, is created by using the process employed for ϕ, but applied in the closed interval $[a, b]$.

For each n and m in $\mathcal{N}$, the interval $[0,1]$ may be divided into 2^n "dyadic" closed subintervals of the form $[m2^{-n}, (m+1)2^{-n}]$, $0 \le m \le 2^n - 1$, generically denoted $[a_k, b_k]$, $1 \le k \le 2^n$. If $0 \le x < y \le 1$, then for some m and n as described, $[a_k, b_k] \subset (x, y)$. Hence

$$\psi \left(\frac{x - a_k}{b_k - a_k} \right) = 0$$

and, since $y > a_k$, $\psi \left(\frac{y - a_k}{b_k - a_k} \right) > 0$ $\left(= \psi \left(\frac{x - a_k}{b_k - a_k} \right) \right)$. Since ψ is a monotone function, for any p in $\mathcal{N}$,

$$\psi \left(\frac{x - a_p}{b_p - a_p} \right) \le \psi \left(\frac{y - a_p}{b_p - a_p} \right)$$

and it follows that $f(x) < f(y)$. The countable set $\mathcal{E}$ consisting of all a_k and b_k is a null set. On the set $U \overset{\text{def}}{=} \{[0, 1] \setminus \mathcal{E}\}$ (a union of open intervals) the function f is locally constant, whence f' exists and is zero.

The argument offered in the footnote of the translation is invalid because the function f is continuous, whence the two superior limits

$$\limsup_{x \to \frac{1}{2}+0} f(x) \text{ and } \limsup_{x \to \frac{1}{2}-0} f(x)$$

are equal.

A CATALOG OF SELECTED
DOVER BOOKS
IN SCIENCE AND MATHEMATICS

Astronomy

BURNHAM'S CELESTIAL HANDBOOK, Robert Burnham, Jr. Thorough guide to the stars beyond our solar system. Exhaustive treatment. Alphabetical by constellation: Andromeda to Cetus in Vol. 1; Chamaeleon to Orion in Vol. 2; and Pavo to Vulpecula in Vol. 3. Hundreds of illustrations. Index in Vol. 3. 2,000pp. 6⅛ x 9¼.

Vol. I: 0-486-23567-X
Vol. II: 0-486-23568-8
Vol. III: 0-486-23673-0

EXPLORING THE MOON THROUGH BINOCULARS AND SMALL TELESCOPES, Ernest H. Cherrington, Jr. Informative, profusely illustrated guide to locating and identifying craters, rills, seas, mountains, other lunar features. Newly revised and updated with special section of new photos. Over 100 photos and diagrams. 240pp. 8¼ x 11. 0-486-24491-1

THE EXTRATERRESTRIAL LIFE DEBATE, 1750–1900, Michael J. Crowe. First detailed, scholarly study in English of the many ideas that developed from 1750 to 1900 regarding the existence of intelligent extraterrestrial life. Examines ideas of Kant, Herschel, Voltaire, Percival Lowell, many other scientists and thinkers. 16 illustrations. 704pp. 5⅜ x 8½. 0-486-40675-X

THEORIES OF THE WORLD FROM ANTIQUITY TO THE COPERNICAN REVOLUTION, Michael J. Crowe. Newly revised edition of an accessible, enlightening book re-creates the change from an earth-centered to a sun-centered conception of the solar system. 242pp. 5⅜ x 8½. 0-486-41444-2

ARISTARCHUS OF SAMOS: The Ancient Copernicus, Sir Thomas Heath. Heath's history of astronomy ranges from Homer and Hesiod to Aristarchus and includes quotes from numerous thinkers, compilers, and scholasticists from Thales and Anaximander through Pythagoras, Plato, Aristotle, and Heraclides. 34 figures. 448pp. 5⅜ x 8½. 0-486-43886-4

A COMPLETE MANUAL OF AMATEUR ASTRONOMY: TOOLS AND TECHNIQUES FOR ASTRONOMICAL OBSERVATIONS, P. Clay Sherrod with Thomas L. Koed. Concise, highly readable book discusses: selecting, setting up and maintaining a telescope; amateur studies of the sun; lunar topography and occultations; observations of Mars, Jupiter, Saturn, the minor planets and the stars; an introduction to photoelectric photometry; more. 1981 ed. 124 figures. 25 halftones. 37 tables. 335pp. 6½ x 9¼. 0-486-42820-8

AMATEUR ASTRONOMER'S HANDBOOK, J. B. Sidgwick. Timeless, comprehensive coverage of telescopes, mirrors, lenses, mountings, telescope drives, micrometers, spectroscopes, more. 189 illustrations. 576pp. 5⅝ x 8¼. (Available in U.S. only.) 0-486-24034-7

STAR LORE: Myths, Legends, and Facts, William Tyler Olcott. Captivating retellings of the origins and histories of ancient star groups include Pegasus, Ursa Major, Pleiades, signs of the zodiac, and other constellations. "Classic."–Sky & Telescope. 58 illustrations. 544pp. 5⅜ x 8½. 0-486-43581-4

Chemistry

THE SCEPTICAL CHYMIST: THE CLASSIC 1661 TEXT, Robert Boyle. Boyle defines the term "element," asserting that all natural phenomena can be explained by the motion and organization of primary particles. 1911 ed. viii+232pp. 5⅜ x 8½.
0-486-42825-7

RADIOACTIVE SUBSTANCES, Marie Curie. Here is the celebrated scientist's doctoral thesis, the prelude to her receipt of the 1903 Nobel Prize. Curie discusses establishing atomic character of radioactivity found in compounds of uranium and thorium; extraction from pitchblende of polonium and radium; isolation of pure radium chloride; determination of atomic weight of radium; plus electric, photographic, luminous, heat, color effects of radioactivity. ii+94pp. 5⅜ x 8½.
0-486-42550-9

CHEMICAL MAGIC, Leonard A. Ford. Second Edition, Revised by E. Winston Grundmeier. Over 100 unusual stunts demonstrating cold fire, dust explosions, much more. Text explains scientific principles and stresses safety precautions. 128pp. 5⅜ x 8½.
0-486-67628-5

MOLECULAR THEORY OF CAPILLARITY, J. S. Rowlinson and B. Widom. History of surface phenomena offers critical and detailed examination and assessment of modern theories, focusing on statistical mechanics and application of results in mean-field approximation to model systems. 1989 edition. 352pp. 5⅜ x 8½.
0-486-42544-4

CHEMICAL AND CATALYTIC REACTION ENGINEERING, James J. Carberry. Designed to offer background for managing chemical reactions, this text examines behavior of chemical reactions and reactors; fluid-fluid and fluid-solid reaction systems; heterogeneous catalysis and catalytic kinetics; more. 1976 edition. 672pp. 6⅛ x 9¼.
0-486-41736-0 $31.95

ELEMENTS OF CHEMISTRY, Antoine Lavoisier. Monumental classic by founder of modern chemistry in remarkable reprint of rare 1790 Kerr translation. A must for every student of chemistry or the history of science. 539pp. 5⅜ x 8½. 0-486-64624-6

MOLECULES AND RADIATION: An Introduction to Modern Molecular Spectroscopy. Second Edition, Jeffrey I. Steinfeld. This unified treatment introduces upper-level undergraduates and graduate students to the concepts and the methods of molecular spectroscopy and applications to quantum electronics, lasers, and related optical phenomena. 1985 edition. 512pp. 5⅜ x 8½. 0-486-44152-0

A SHORT HISTORY OF CHEMISTRY, J. R. Partington. Classic exposition explores origins of chemistry, alchemy, early medical chemistry, nature of atmosphere, theory of valency, laws and structure of atomic theory, much more. 428pp. 5⅜ x 8½. (Available in U.S. only.) 0-486-65977-1

GENERAL CHEMISTRY, Linus Pauling. Revised 3rd edition of classic first-year text by Nobel laureate. Atomic and molecular structure, quantum mechanics, statistical mechanics, thermodynamics correlated with descriptive chemistry. Problems. 992pp. 5⅜ x 8½. 0-486-65622-5

ELECTRON CORRELATION IN MOLECULES, S. Wilson. This text addresses one of theoretical chemistry's central problems. Topics include molecular electronic structure, independent electron models, electron correlation, the linked diagram theorem, and related topics. 1984 edition. 304pp. 5⅜ x 8½. 0-486-45879-2

Engineering

DE RE METALLICA, Georgius Agricola. The famous Hoover translation of greatest treatise on technological chemistry, engineering, geology, mining of early modern times (1556). All 289 original woodcuts. 638pp. 6¾ x 11.   0-486-60006-8

FUNDAMENTALS OF ASTRODYNAMICS, Roger Bate et al. Modern approach developed by U.S. Air Force Academy. Designed as a first course. Problems, exercises. Numerous illustrations. 455pp. 5⅜ x 8½.   0-486-60061-0

DYNAMICS OF FLUIDS IN POROUS MEDIA, Jacob Bear. For advanced students of ground water hydrology, soil mechanics and physics, drainage and irrigation engineering and more. 335 illustrations. Exercises, with answers. 784pp. 6⅛ x 9¼.
0-486-65675-6

THEORY OF VISCOELASTICITY (SECOND EDITION), Richard M. Christensen. Complete consistent description of the linear theory of the viscoelastic behavior of materials. Problem-solving techniques discussed. 1982 edition. 29 figures. xiv+364pp. 6⅛ x 9¼.   0-486-42880-X

MECHANICS, J. P. Den Hartog. A classic introductory text or refresher. Hundreds of applications and design problems illuminate fundamentals of trusses, loaded beams and cables, etc. 334 answered problems. 462pp. 5⅜ x 8½.   0-486-60754-2

MECHANICAL VIBRATIONS, J. P. Den Hartog. Classic textbook offers lucid explanations and illustrative models, applying theories of vibrations to a variety of practical industrial engineering problems. Numerous figures. 233 problems, solutions. Appendix. Index. Preface. 436pp. 5⅜ x 8½.   0-486-64785-4

STRENGTH OF MATERIALS, J. P. Den Hartog. Full, clear treatment of basic material (tension, torsion, bending, etc.) plus advanced material on engineering methods, applications. 350 answered problems. 323pp. 5⅜ x 8½.   0-486-60755-0

A HISTORY OF MECHANICS, René Dugas. Monumental study of mechanical principles from antiquity to quantum mechanics. Contributions of ancient Greeks, Galileo, Leonardo, Kepler, Lagrange, many others. 671pp. 5⅜ x 8½. 0-486-65632-2

STABILITY THEORY AND ITS APPLICATIONS TO STRUCTURAL MECHANICS, Clive L. Dym. Self-contained text focuses on Koiter postbuckling analyses, with mathematical notions of stability of motion. Basing minimum energy principles for static stability upon dynamic concepts of stability of motion, it develops asymptotic buckling and postbuckling analyses from potential energy considerations, with applications to columns, plates, and arches. 1974 ed. 208pp. 5⅜ x 8½.
0-486-42541-X

BASIC ELECTRICITY, U.S. Bureau of Naval Personnel. Originally a training course; best nontechnical coverage. Topics include batteries, circuits, conductors, AC and DC, inductance and capacitance, generators, motors, transformers, amplifiers, etc. Many questions with answers. 349 illustrations. 1969 edition. 448pp. 6½ x 9¼.
0-486-20973-3

ROCKETS, Robert Goddard. Two of the most significant publications in the history of rocketry and jet propulsion: "A Method of Reaching Extreme Altitudes" (1919) and "Liquid Propellant Rocket Development" (1936). 128pp. 5⅜ x 8½.   0-486-42537-1

STATISTICAL MECHANICS: PRINCIPLES AND APPLICATIONS, Terrell L. Hill. Standard text covers fundamentals of statistical mechanics, applications to fluctuation theory, imperfect gases, distribution functions, more. 448pp. 5⅜ x 8½.
0-486-65390-0

ENGINEERING AND TECHNOLOGY 1650–1750: ILLUSTRATIONS AND TEXTS FROM ORIGINAL SOURCES, Martin Jensen. Highly readable text with more than 200 contemporary drawings and detailed engravings of engineering projects dealing with surveying, leveling, materials, hand tools, lifting equipment, transport and erection, piling, bailing, water supply, hydraulic engineering, and more. Among the specific projects outlined-transporting a 50-ton stone to the Louvre, erecting an obelisk, building timber locks, and dredging canals. 207pp. 8⅜ x 11¼.
0-486-42232-1

THE VARIATIONAL PRINCIPLES OF MECHANICS, Cornelius Lanczos. Graduate level coverage of calculus of variations, equations of motion, relativistic mechanics, more. First inexpensive paperbound edition of classic treatise. Index. Bibliography. 418pp. 5⅜ x 8½.   0-486-65067-7

PROTECTION OF ELECTRONIC CIRCUITS FROM OVERVOLTAGES, Ronald B. Standler. Five-part treatment presents practical rules and strategies for circuits designed to protect electronic systems from damage by transient overvoltages. 1989 ed. xxiv+434pp. 6⅛ x 9¼.   0-486-42552-5

ROTARY WING AERODYNAMICS, W. Z. Stepniewski. Clear, concise text covers aerodynamic phenomena of the rotor and offers guidelines for helicopter performance evaluation. Originally prepared for NASA. 537 figures. 640pp. 6⅛ x 9¼.
0-486-64647-5

INTRODUCTION TO SPACE DYNAMICS, William Tyrrell Thomson. Comprehensive, classic introduction to space-flight engineering for advanced undergraduate and graduate students. Includes vector algebra, kinematics, transformation of coordinates. Bibliography. Index. 352pp. 5⅜ x 8½.   0-486-65113-4

HISTORY OF STRENGTH OF MATERIALS, Stephen P. Timoshenko. Excellent historical survey of the strength of materials with many references to the theories of elasticity and structure. 245 figures. 452pp. 5⅜ x 8½.   0-486-61187-6

ANALYTICAL FRACTURE MECHANICS, David J. Unger. Self-contained text supplements standard fracture mechanics texts by focusing on analytical methods for determining crack-tip stress and strain fields. 336pp. 6⅛ x 9¼.   0-486-41737-9

STATISTICAL MECHANICS OF ELASTICITY, J. H. Weiner. Advanced, self-contained treatment illustrates general principles and elastic behavior of solids. Part 1, based on classical mechanics, studies thermoelastic behavior of crystalline and polymeric solids. Part 2, based on quantum mechanics, focuses on interatomic force laws, behavior of solids, and thermally activated processes. For students of physics and chemistry and for polymer physicists. 1983 ed. 96 figures. 496pp. 5⅜ x 8½.
0-486-42260-7

Mathematics

FUNCTIONAL ANALYSIS (Second Corrected Edition), George Bachman and Lawrence Narici. Excellent treatment of subject geared toward students with background in linear algebra, advanced calculus, physics and engineering. Text covers introduction to inner-product spaces, normed, metric spaces, and topological spaces; complete orthonormal sets, the Hahn-Banach Theorem and its consequences, and many other related subjects. 1966 ed. 544pp. 6⅛ x 9¼.   0-486-40251-7

DIFFERENTIAL MANIFOLDS, Antoni A. Kosinski. Introductory text for advanced undergraduates and graduate students presents systematic study of the topological structure of smooth manifolds, starting with elements of theory and concluding with method of surgery. 1993 edition. 288pp. 5⅜ x 8½.   0-486-46244-7

VECTOR AND TENSOR ANALYSIS WITH APPLICATIONS, A. I. Borisenko and I. E. Tarapov. Concise introduction. Worked-out problems, solutions, exercises. 257pp. 5⅝ x 8¼.   0-486-63833-2

AN INTRODUCTION TO ORDINARY DIFFERENTIAL EQUATIONS, Earl A. Coddington. A thorough and systematic first course in elementary differential equations for undergraduates in mathematics and science, with many exercises and problems (with answers). Index. 304pp. 5⅜ x 8½.   0-486-65942-9

FOURIER SERIES AND ORTHOGONAL FUNCTIONS, Harry F. Davis. An incisive text combining theory and practical example to introduce Fourier series, orthogonal functions and applications of the Fourier method to boundary-value problems. 570 exercises. Answers and notes. 416pp. 5⅜ x 8½.   0-486-65973-9

COMPUTABILITY AND UNSOLVABILITY, Martin Davis. Classic graduate-level introduction to theory of computability, usually referred to as theory of recurrent functions. New preface and appendix. 288pp. 5⅜ x 8½.   0-486-61471-9

AN INTRODUCTION TO MATHEMATICAL ANALYSIS, Robert A. Rankin. Dealing chiefly with functions of a single real variable, this text by a distinguished educator introduces limits, continuity, differentiability, integration, convergence of infinite series, double series, and infinite products. 1963 edition. 624pp. 5⅜ x 8½.
0-486-46251-X

METHODS OF NUMERICAL INTEGRATION (SECOND EDITION), Philip J. Davis and Philip Rabinowitz. Requiring only a background in calculus, this text covers approximate integration over finite and infinite intervals, error analysis, approximate integration in two or more dimensions, and automatic integration. 1984 edition. 624pp. 5⅜ x 8½.   0-486-45339-1

INTRODUCTION TO LINEAR ALGEBRA AND DIFFERENTIAL EQUATIONS, John W. Dettman. Excellent text covers complex numbers, determinants, orthonormal bases, Laplace transforms, much more. Exercises with solutions. Undergraduate level. 416pp. 5⅜ x 8½.   0-486-65191-6

RIEMANN'S ZETA FUNCTION, H. M. Edwards. Superb, high-level study of landmark 1859 publication entitled "On the Number of Primes Less Than a Given Magnitude" traces developments in mathematical theory that it inspired. xiv+315pp. 5⅜ x 8½.   0-486-41740-9

CALCULUS OF VARIATIONS WITH APPLICATIONS, George M. Ewing. Applications-oriented introduction to variational theory develops insight and promotes understanding of specialized books, research papers. Suitable for advanced undergraduate/graduate students as primary, supplementary text. 352pp. 5⅜ x 8½.
0-486-64856-7

MATHEMATICIAN'S DELIGHT, W. W. Sawyer. "Recommended with confidence" by *The Times Literary Supplement,* this lively survey was written by a renowned teacher. It starts with arithmetic and algebra, gradually proceeding to trigonometry and calculus. 1943 edition. 240pp. 5⅜ x 8½.
0-486-46240-4

ADVANCED EUCLIDEAN GEOMETRY, Roger A. Johnson. This classic text explores the geometry of the triangle and the circle, concentrating on extensions of Euclidean theory, and examining in detail many relatively recent theorems. 1929 edition. 336pp. 5⅜ x 8½.
0-486-46237-4

COUNTEREXAMPLES IN ANALYSIS, Bernard R. Gelbaum and John M. H. Olmsted. These counterexamples deal mostly with the part of analysis known as "real variables." The first half covers the real number system, and the second half encompasses higher dimensions. 1962 edition. xxiv+198pp. 5⅜ x 8½. 0-486-42875-3

CATASTROPHE THEORY FOR SCIENTISTS AND ENGINEERS, Robert Gilmore. Advanced-level treatment describes mathematics of theory grounded in the work of Poincaré, R. Thom, other mathematicians. Also important applications to problems in mathematics, physics, chemistry and engineering. 1981 edition. References. 28 tables. 397 black-and-white illustrations. xvii + 666pp. 6⅛ x 9¼.
0-486-67539-4

COMPLEX VARIABLES: Second Edition, Robert B. Ash and W. P. Novinger. Suitable for advanced undergraduates and graduate students, this newly revised treatment covers Cauchy theorem and its applications, analytic functions, and the prime number theorem. Numerous problems and solutions. 2004 edition. 224pp. 6½ x 9¼.
0-486-46250-1

NUMERICAL METHODS FOR SCIENTISTS AND ENGINEERS, Richard Hamming. Classic text stresses frequency approach in coverage of algorithms, polynomial approximation, Fourier approximation, exponential approximation, other topics. Revised and enlarged 2nd edition. 721pp. 5⅜ x 8½.
0-486-65241-6

INTRODUCTION TO NUMERICAL ANALYSIS (2nd Edition), F. B. Hildebrand. Classic, fundamental treatment covers computation, approximation, interpolation, numerical differentiation and integration, other topics. 150 new problems. 669pp. 5⅜ x 8½.
0-486-65363-3

MARKOV PROCESSES AND POTENTIAL THEORY, Robert M. Blumental and Ronald K. Getoor. This graduate-level text explores the relationship between Markov processes and potential theory in terms of excessive functions, multiplicative functionals and subprocesses, additive functionals and their potentials, and dual processes. 1968 edition. 320pp. 5⅜ x 8½.
0-486-46263-3

ABSTRACT SETS AND FINITE ORDINALS: An Introduction to the Study of Set Theory, G. B. Keene. This text unites logical and philosophical aspects of set theory in a manner intelligible to mathematicians without training in formal logic and to logicians without a mathematical background. 1961 edition. 112pp. 5⅜ x 8½.
0-486-46249-8

INTRODUCTORY REAL ANALYSIS, A.N. Kolmogorov, S. V. Fomin. Translated by Richard A. Silverman. Self-contained, evenly paced introduction to real and functional analysis. Some 350 problems. 403pp. 5⅜ x 8½.   0-486-61226-0

APPLIED ANALYSIS, Cornelius Lanczos. Classic work on analysis and design of finite processes for approximating solution of analytical problems. Algebraic equations, matrices, harmonic analysis, quadrature methods, much more. 559pp. 5⅜ x 8½.   0-486-65656-X

AN INTRODUCTION TO ALGEBRAIC STRUCTURES, Joseph Landin. Superb self-contained text covers "abstract algebra": sets and numbers, theory of groups, theory of rings, much more. Numerous well-chosen examples, exercises. 247pp. 5⅜ x 8½.   0-486-65940-2

QUALITATIVE THEORY OF DIFFERENTIAL EQUATIONS, V. V. Nemytskii and V.V. Stepanov. Classic graduate-level text by two prominent Soviet mathematicians covers classical differential equations as well as topological dynamics and ergodic theory. Bibliographies. 523pp. 5⅜ x 8½.   0-486-65954-2

THEORY OF MATRICES, Sam Perlis. Outstanding text covering rank, nonsingularity and inverses in connection with the development of canonical matrices under the relation of equivalence, and without the intervention of determinants. Includes exercises. 237pp. 5⅜ x 8½.   0-486-66810-X

INTRODUCTION TO ANALYSIS, Maxwell Rosenlicht. Unusually clear, accessible coverage of set theory, real number system, metric spaces, continuous functions, Riemann integration, multiple integrals, more. Wide range of problems. Undergraduate level. Bibliography. 254pp. 5⅜ x 8½.   0-486-65038-3

MODERN NONLINEAR EQUATIONS, Thomas L. Saaty. Emphasizes practical solution of problems; covers seven types of equations. ". . . a welcome contribution to the existing literature. . . ."–*Math Reviews.* 490pp. 5⅜ x 8½.   0-486-64232-1

MATRICES AND LINEAR ALGEBRA, Hans Schneider and George Phillip Barker. Basic textbook covers theory of matrices and its applications to systems of linear equations and related topics such as determinants, eigenvalues and differential equations. Numerous exercises. 432pp. 5⅜ x 8½.   0-486-66014-1

LINEAR ALGEBRA, Georgi E. Shilov. Determinants, linear spaces, matrix algebras, similar topics. For advanced undergraduates, graduates. Silverman translation. 387pp. 5⅜ x 8½.   0-486-63518-X

MATHEMATICAL METHODS OF GAME AND ECONOMIC THEORY: Revised Edition, Jean-Pierre Aubin. This text begins with optimization theory and convex analysis, followed by topics in game theory and mathematical economics, and concluding with an introduction to nonlinear analysis and control theory. 1982 edition. 656pp. 6⅛ x 9¼.   0-486-46265-X

SET THEORY AND LOGIC, Robert R. Stoll. Lucid introduction to unified theory of mathematical concepts. Set theory and logic seen as tools for conceptual understanding of real number system. 496pp. 5⅝ x 8¼.   0-486-63829-4

CATALOG OF DOVER BOOKS

TENSOR CALCULUS, J.L. Synge and A. Schild. Widely used introductory text covers spaces and tensors, basic operations in Riemannian space, non-Riemannian spaces, etc. 324pp. 5⅝ x 8¼.   0-486-63612-7

ORDINARY DIFFERENTIAL EQUATIONS, Morris Tenenbaum and Harry Pollard. Exhaustive survey of ordinary differential equations for undergraduates in mathematics, engineering, science. Thorough analysis of theorems. Diagrams. Bibliography. Index. 818pp. 5⅜ x 8½.   0-486-64940-7

INTEGRAL EQUATIONS, F. G. Tricomi. Authoritative, well-written treatment of extremely useful mathematical tool with wide applications. Volterra Equations, Fredholm Equations, much more. Advanced undergraduate to graduate level. Exercises. Bibliography. 238pp. 5⅜ x 8½.   0-486-64828-1

FOURIER SERIES, Georgi P. Tolstov. Translated by Richard A. Silverman. A valuable addition to the literature on the subject, moving clearly from subject to subject and theorem to theorem. 107 problems, answers. 336pp. 5⅜ x 8½.   0-486-63317-9

INTRODUCTION TO MATHEMATICAL THINKING, Friedrich Waismann. Examinations of arithmetic, geometry, and theory of integers; rational and natural numbers; complete induction; limit and point of accumulation; remarkable curves; complex and hypercomplex numbers, more. 1959 ed. 27 figures. xii+260pp. 5⅜ x 8½.   0-486-42804-8

THE RADON TRANSFORM AND SOME OF ITS APPLICATIONS, Stanley R. Deans. Of value to mathematicians, physicists, and engineers, this excellent introduction covers both theory and applications, including a rich array of examples and literature. Revised and updated by the author. 1993 edition. 304pp. 6⅛ x 9¼.   0-486-46241-2

CALCULUS OF VARIATIONS, Robert Weinstock. Basic introduction covering isoperimetric problems, theory of elasticity, quantum mechanics, electrostatics, etc. Exercises throughout. 326pp. 5⅜ x 8½.   0-486-63069-2

THE CONTINUUM: A CRITICAL EXAMINATION OF THE FOUNDATION OF ANALYSIS, Hermann Weyl. Classic of 20th-century foundational research deals with the conceptual problem posed by the continuum. 156pp. 5⅜ x 8½.   0-486-67982-9

CHALLENGING MATHEMATICAL PROBLEMS WITH ELEMENTARY SOLUTIONS, A. M. Yaglom and I. M. Yaglom. Over 170 challenging problems on probability theory, combinatorial analysis, points and lines, topology, convex polygons, many other topics. Solutions. Total of 445pp. 5⅜ x 8½. Two-vol. set.
          Vol. I: 0-486-65536-9   Vol. II: 0-486-65537-7

INTRODUCTION TO PARTIAL DIFFERENTIAL EQUATIONS WITH APPLICATIONS, E. C. Zachmanoglou and Dale W. Thoe. Essentials of partial differential equations applied to common problems in engineering and the physical sciences. Problems and answers. 416pp. 5⅜ x 8½.   0-486-65251-3

STOCHASTIC PROCESSES AND FILTERING THEORY, Andrew H. Jazwinski. This unified treatment presents material previously available only in journals, and in terms accessible to engineering students. Although theory is emphasized, it discusses numerous practical applications as well. 1970 edition. 400pp. 5⅜ x 8½. 0-486-46274-9

Math–Decision Theory, Statistics, Probability

INTRODUCTION TO PROBABILITY, John E. Freund. Featured topics include permutations and factorials, probabilities and odds, frequency interpretation, mathematical expectation, decision-making, postulates of probability, rule of elimination, much more. Exercises with some solutions. Summary. 1973 edition. 247pp. 5⅜ x 8½.
0-486-67549-1

STATISTICAL AND INDUCTIVE PROBABILITIES, Hugues Leblanc. This treatment addresses a decades-old dispute among probability theorists, asserting that both statistical and inductive probabilities may be treated as sentence-theoretic measurements, and that the latter qualify as estimates of the former. 1962 edition. 160pp. 5⅜ x 8½.
0-486-44980-7

APPLIED MULTIVARIATE ANALYSIS: Using Bayesian and Frequentist Methods of Inference, Second Edition, S. James Press. This two-part treatment deals with foundations as well as models and applications. Topics include continuous multivariate distributions; regression and analysis of variance; factor analysis and latent structure analysis; and structuring multivariate populations. 1982 edition. 692pp. 5⅜ x 8½.
0-486-44236-5

LINEAR PROGRAMMING AND ECONOMIC ANALYSIS, Robert Dorfman, Paul A. Samuelson and Robert M. Solow. First comprehensive treatment of linear programming in standard economic analysis. Game theory, modern welfare economics, Leontief input-output, more. 525pp. 5⅜ x 8½.
0-486-65491-5

PROBABILITY: AN INTRODUCTION, Samuel Goldberg. Excellent basic text covers set theory, probability theory for finite sample spaces, binomial theorem, much more. 360 problems. Bibliographies. 322pp. 5⅜ x 8½.
0-486-65252-1

GAMES AND DECISIONS: INTRODUCTION AND CRITICAL SURVEY, R. Duncan Luce and Howard Raiffa. Superb nontechnical introduction to game theory, primarily applied to social sciences. Utility theory, zero-sum games, n-person games, decision-making, much more. Bibliography. 509pp. 5⅜ x 8½. 0-486-65943-7

INTRODUCTION TO THE THEORY OF GAMES, J. C. C. McKinsey. This comprehensive overview of the mathematical theory of games illustrates applications to situations involving conflicts of interest, including economic, social, political, and military contexts. Appropriate for advanced undergraduate and graduate courses; advanced calculus a prerequisite. 1952 ed. x+372pp. 5⅜ x 8½.   0-486-42811-7

FIFTY CHALLENGING PROBLEMS IN PROBABILITY WITH SOLUTIONS, Frederick Mosteller. Remarkable puzzlers, graded in difficulty, illustrate elementary and advanced aspects of probability. Detailed solutions. 88pp. 5⅜ x 8½.
0-486-65355-2

PROBABILITY THEORY: A CONCISE COURSE, Y. A. Rozanov. Highly readable, self-contained introduction covers combination of events, dependent events, Bernoulli trials, etc. 148pp. 5⅝ x 8¼.   0-486-63544-9

THE STATISTICAL ANALYSIS OF EXPERIMENTAL DATA, John Mandel. First half of book presents fundamental mathematical definitions, concepts and facts while remaining half deals with statistics primarily as an interpretive tool. Well-written text, numerous worked examples with step-by-step presentation. Includes 116 tables. 448pp. 5⅜ x 8½.   0-486-64666-1

Math–Geometry and Topology

ELEMENTARY CONCEPTS OF TOPOLOGY, Paul Alexandroff. Elegant, intuitive approach to topology from set-theoretic topology to Betti groups; how concepts of topology are useful in math and physics. 25 figures. 57pp. 5³/₈ x 8¹/₂. 0-486-60747-X

A LONG WAY FROM EUCLID, Constance Reid. Lively guide by a prominent historian focuses on the role of Euclid's Elements in subsequent mathematical developments. Elementary algebra and plane geometry are sole prerequisites. 80 drawings. 1963 edition. 304pp. 5³/₈ x 8¹/₂. 0-486-43613-6

EXPERIMENTS IN TOPOLOGY, Stephen Barr. Classic, lively explanation of one of the byways of mathematics. Klein bottles, Moebius strips, projective planes, map coloring, problem of the Koenigsberg bridges, much more, described with clarity and wit. 43 figures. 210pp. 5³/₈ x 8¹/₂. 0-486-25933-1

THE GEOMETRY OF RENÉ DESCARTES, René Descartes. The great work founded analytical geometry. Original French text, Descartes's own diagrams, together with definitive Smith-Latham translation. 244pp. 5³/₈ x 8¹/₂. 0-486-60068-8

EUCLIDEAN GEOMETRY AND TRANSFORMATIONS, Clayton W. Dodge. This introduction to Euclidean geometry emphasizes transformations, particularly isometries and similarities. Suitable for undergraduate courses, it includes numerous examples, many with detailed answers. 1972 ed. viii+296pp. 6¹/₈ x 9¹/₄. 0-486-43476-1

EXCURSIONS IN GEOMETRY, C. Stanley Ogilvy. A straightedge, compass, and a little thought are all that's needed to discover the intellectual excitement of geometry. Harmonic division and Apollonian circles, inversive geometry, hexlet, Golden Section, more. 132 illustrations. 192pp. 5³/₈ x 8¹/₂. 0-486-26530-7

THE THIRTEEN BOOKS OF EUCLID'S ELEMENTS, translated with introduction and commentary by Sir Thomas L. Heath. Definitive edition. Textual and linguistic notes, mathematical analysis. 2,500 years of critical commentary. Unabridged. 1,414pp. 5³/₈ x 8¹/₂. Three-vol. set.
Vol. I: 0-486-60088-2 Vol. II: 0-486-60089-0 Vol. III: 0-486-60090-4

SPACE AND GEOMETRY: IN THE LIGHT OF PHYSIOLOGICAL, PSYCHOLOGICAL AND PHYSICAL INQUIRY, Ernst Mach. Three essays by an eminent philosopher and scientist explore the nature, origin, and development of our concepts of space, with a distinctness and precision suitable for undergraduate students and other readers. 1906 ed. vi+148pp. 5³/₈ x 8¹/₂. 0-486-43909-7

GEOMETRY OF COMPLEX NUMBERS, Hans Schwerdtfeger. Illuminating, widely praised book on analytic geometry of circles, the Moebius transformation, and two-dimensional non-Euclidean geometries. 200pp. 5⁵/₈ x 8¹/₄. 0-486-63830-8

DIFFERENTIAL GEOMETRY, Heinrich W. Guggenheimer. Local differential geometry as an application of advanced calculus and linear algebra. Curvature, transformation groups, surfaces, more. Exercises. 62 figures. 378pp. 5³/₈ x 8¹/₂. 0-486-63433-7

History of Math

THE WORKS OF ARCHIMEDES, Archimedes (T. L. Heath, ed.). Topics include the famous problems of the ratio of the areas of a cylinder and an inscribed sphere; the measurement of a circle; the properties of conoids, spheroids, and spirals; and the quadrature of the parabola. Informative introduction. clxxxvi+326pp. 5⅜ x 8½.
0-486-42084-1

A SHORT ACCOUNT OF THE HISTORY OF MATHEMATICS, W. W. Rouse Ball. One of clearest, most authoritative surveys from the Egyptians and Phoenicians through 19th-century figures such as Grassman, Galois, Riemann. Fourth edition. 522pp. 5⅜ x 8½.
0-486-20630-0

THE HISTORY OF THE CALCULUS AND ITS CONCEPTUAL DEVELOP-MENT, Carl B. Boyer. Origins in antiquity, medieval contributions, work of Newton, Leibniz, rigorous formulation. Treatment is verbal. 346pp. 5⅜ x 8½. 0-486-60509-4

THE HISTORICAL ROOTS OF ELEMENTARY MATHEMATICS, Lucas N. H. Bunt, Phillip S. Jones, and Jack D. Bedient. Fundamental underpinnings of modern arithmetic, algebra, geometry and number systems derived from ancient civilizations. 320pp. 5⅜ x 8½.
0-486-25563-8

THE HISTORY OF THE CALCULUS AND ITS CONCEPTUAL DEVELOP-MENT, Carl B. Boyer. Fluent description of the development of both the integral and differential calculus—its early beginnings in antiquity, medieval contributions, and a consideration of Newton and Leibniz. 368pp. 5⅜ x 8½. 0-486-60509-4

GAMES, GODS & GAMBLING: A HISTORY OF PROBABILITY AND STATISTICAL IDEAS, F. N. David. Episodes from the lives of Galileo, Fermat, Pascal, and others illustrate this fascinating account of the roots of mathematics. Features thought-provoking references to classics, archaeology, biography, poetry. 1962 edition. 304pp. 5⅜ x 8½. (Available in U.S. only.)
0-486-40023-9

OF MEN AND NUMBERS: THE STORY OF THE GREAT MATHEMATICIANS, Jane Muir. Fascinating accounts of the lives and accomplishments of history's greatest mathematical minds—Pythagoras, Descartes, Euler, Pascal, Cantor, many more. Anecdotal, illuminating. 30 diagrams. Bibliography. 256pp. 5⅜ x 8½.
0-486-28973-7

HISTORY OF MATHEMATICS, David E. Smith. Nontechnical survey from ancient Greece and Orient to late 19th century; evolution of arithmetic, geometry, trigonometry, calculating devices, algebra, the calculus. 362 illustrations. 1,355pp. 5⅜ x 8½. Two-vol. set.   Vol. I: 0-486-20429-4   Vol. II: 0-486-20430-8

A CONCISE HISTORY OF MATHEMATICS, Dirk J. Struik. The best brief history of mathematics. Stresses origins and covers every major figure from ancient Near East to 19th century. 41 illustrations. 195pp. 5⅜ x 8½.   0-486-60255-9

Physics

OPTICAL RESONANCE AND TWO-LEVEL ATOMS, L. Allen and J. H. Eberly. Clear, comprehensive introduction to basic principles behind all quantum optical resonance phenomena. 53 illustrations. Preface. Index. 256pp. 5⅜ x 8½.
0-486-65533-4

QUANTUM THEORY, David Bohm. This advanced undergraduate-level text presents the quantum theory in terms of qualitative and imaginative concepts, followed by specific applications worked out in mathematical detail. Preface. Index. 655pp. 5⅜ x 8½.
0-486-65969-0

ATOMIC PHYSICS (8th EDITION), Max Born. Nobel laureate's lucid treatment of kinetic theory of gases, elementary particles, nuclear atom, wave-corpuscles, atomic structure and spectral lines, much more. Over 40 appendices, bibliography. 495pp. 5⅜ x 8½.
0-486-65984-4

A SOPHISTICATE'S PRIMER OF RELATIVITY, P. W. Bridgman. Geared toward readers already acquainted with special relativity, this book transcends the view of theory as a working tool to answer natural questions: What is a frame of reference? What is a "law of nature"? What is the role of the "observer"? Extensive treatment, written in terms accessible to those without a scientific background. 1983 ed. xlviii+172pp. 5⅜ x 8½.
0-486-42549-5

AN INTRODUCTION TO HAMILTONIAN OPTICS, H. A. Buchdahl. Detailed account of the Hamiltonian treatment of aberration theory in geometrical optics. Many classes of optical systems defined in terms of the symmetries they possess. Problems with detailed solutions. 1970 edition. xv + 360pp. 5⅜ x 8½. 0-486-67597-1

PRIMER OF QUANTUM MECHANICS, Marvin Chester. Introductory text examines the classical quantum bead on a track: its state and representations; operator eigenvalues; harmonic oscillator and bound bead in a symmetric force field; and bead in a spherical shell. Other topics include spin, matrices, and the structure of quantum mechanics; the simplest atom; indistinguishable particles; and stationary-state perturbation theory. 1992 ed. xiv+314pp. 6⅛ x 9¼.
0-486-42878-8

LECTURES ON QUANTUM MECHANICS, Paul A. M. Dirac. Four concise, brilliant lectures on mathematical methods in quantum mechanics from Nobel Prize-winning quantum pioneer build on idea of visualizing quantum theory through the use of classical mechanics. 96pp. 5⅜ x 8½.
0-486-41713-1

THIRTY YEARS THAT SHOOK PHYSICS: THE STORY OF QUANTUM THEORY, George Gamow. Lucid, accessible introduction to influential theory of energy and matter. Careful explanations of Dirac's anti-particles, Bohr's model of the atom, much more. 12 plates. Numerous drawings. 240pp. 5⅜ x 8½. 0-486-24895-X

ELECTRONIC STRUCTURE AND THE PROPERTIES OF SOLIDS: THE PHYSICS OF THE CHEMICAL BOND, Walter A. Harrison. Innovative text offers basic understanding of the electronic structure of covalent and ionic solids, simple metals, transition metals and their compounds. Problems. 1980 edition. 582pp. 6⅛ x 9¼.
0-486-66021-4

CATALOG OF DOVER BOOKS

HYDRODYNAMIC AND HYDROMAGNETIC STABILITY, S. Chandrasekhar. Lucid examination of the Rayleigh-Benard problem; clear coverage of the theory of instabilities causing convection. 704pp. 5⅝ x 8¼. 0-486-64071-X

INVESTIGATIONS ON THE THEORY OF THE BROWNIAN MOVEMENT, Albert Einstein. Five papers (1905–8) investigating dynamics of Brownian motion and evolving elementary theory. Notes by R. Fürth. 122pp. 5⅜ x 8½. 0-486-60304-0

THE PHYSICS OF WAVES, William C. Elmore and Mark A. Heald. Unique overview of classical wave theory. Acoustics, optics, electromagnetic radiation, more. Ideal as classroom text or for self-study. Problems. 477pp. 5⅜ x 8½. 0-486-64926-1

GRAVITY, George Gamow. Distinguished physicist and teacher takes reader-friendly look at three scientists whose work unlocked many of the mysteries behind the laws of physics: Galileo, Newton, and Einstein. Most of the book focuses on Newton's ideas, with a concluding chapter on post-Einsteinian speculations concerning the relationship between gravity and other physical phenomena. 160pp. 5⅜ x 8½. 0-486-42563-0

PHYSICAL PRINCIPLES OF THE QUANTUM THEORY, Werner Heisenberg. Nobel Laureate discusses quantum theory, uncertainty, wave mechanics, work of Dirac, Schroedinger, Compton, Wilson, Einstein, etc. 184pp. 5⅜ x 8½. 0-486-60113-7

ATOMIC SPECTRA AND ATOMIC STRUCTURE, Gerhard Herzberg. One of best introductions; especially for specialist in other fields. Treatment is physical rather than mathematical. 80 illustrations. 257pp. 5⅜ x 8½. 0-486-60115-3

AN INTRODUCTION TO STATISTICAL THERMODYNAMICS, Terrell L. Hill. Excellent basic text offers wide-ranging coverage of quantum statistical mechanics, systems of interacting molecules, quantum statistics, more. 523pp. 5⅜ x 8½. 0-486-65242-4

THEORETICAL PHYSICS, Georg Joos, with Ira M. Freeman. Classic overview covers essential math, mechanics, electromagnetic theory, thermodynamics, quantum mechanics, nuclear physics, other topics. First paperback edition. xxiii + 885pp. 5⅜ x 8½. 0-486-65227-0

PROBLEMS AND SOLUTIONS IN QUANTUM CHEMISTRY AND PHYSICS, Charles S. Johnson, Jr. and Lee G. Pedersen. Unusually varied problems, detailed solutions in coverage of quantum mechanics, wave mechanics, angular momentum, molecular spectroscopy, more. 280 problems plus 139 supplementary exercises. 430pp. 6½ x 9¼. 0-486-65236-X

THEORETICAL SOLID STATE PHYSICS, Vol. 1: Perfect Lattices in Equilibrium; Vol. II: Non-Equilibrium and Disorder, William Jones and Norman H. March. Monumental reference work covers fundamental theory of equilibrium properties of perfect crystalline solids, non-equilibrium properties, defects and disordered systems. Appendices. Problems. Preface. Diagrams. Index. Bibliography. Total of 1,301pp. 5⅜ x 8½. Two volumes. Vol. I: 0-486-65015-4   Vol. II: 0-486-65016-2

WHAT IS RELATIVITY? L. D. Landau and G. B. Rumer. Written by a Nobel Prize physicist and his distinguished colleague, this compelling book explains the special theory of relativity to readers with no scientific background, using such familiar objects as trains, rulers, and clocks. 1960 ed. vi+72pp. 5⅜ x 8½. 0-486-42806-0

CATALOG OF DOVER BOOKS

A TREATISE ON ELECTRICITY AND MAGNETISM, James Clerk Maxwell. Important foundation work of modern physics. Brings to final form Maxwell's theory of electromagnetism and rigorously derives his general equations of field theory. 1,084pp. 5⅜ x 8½. Two-vol. set.   Vol. I: 0-486-60636-8   Vol. II: 0-486-60637-6

MATHEMATICS FOR PHYSICISTS, Philippe Dennery and Andre Krzywicki. Superb text provides math needed to understand today's more advanced topics in physics and engineering. Theory of functions of a complex variable, linear vector spaces, much more. Problems. 1967 edition. 400pp. 6½ x 9¼.   0-486-69193-4

INTRODUCTION TO QUANTUM MECHANICS WITH APPLICATIONS TO CHEMISTRY, Linus Pauling & E. Bright Wilson, Jr. Classic undergraduate text by Nobel Prize winner applies quantum mechanics to chemical and physical problems. Numerous tables and figures enhance the text. Chapter bibliographies. Appendices. Index. 468pp. 5⅜ x 8½.   0-486-64871-0

METHODS OF THERMODYNAMICS, Howard Reiss. Outstanding text focuses on physical technique of thermodynamics, typical problem areas of understanding, and significance and use of thermodynamic potential. 1965 edition. 238pp. 5⅜ x 8½.
0-486-69445-3

THE ELECTROMAGNETIC FIELD, Albert Shadowitz. Comprehensive undergraduate text covers basics of electric and magnetic fields, builds up to electromagnetic theory. Also related topics, including relativity. Over 900 problems. 768pp. 5⅜ x 8¼.   0-486-65660-8

GREAT EXPERIMENTS IN PHYSICS: FIRSTHAND ACCOUNTS FROM GALILEO TO EINSTEIN, Morris H. Shamos (ed.). 25 crucial discoveries: Newton's laws of motion, Chadwick's study of the neutron, Hertz on electromagnetic waves, more. Original accounts clearly annotated. 370pp. 5⅜ x 8½.   0-486-25346-5

EINSTEIN'S LEGACY, Julian Schwinger. A Nobel Laureate relates fascinating story of Einstein and development of relativity theory in well-illustrated, nontechnical volume. Subjects include meaning of time, paradoxes of space travel, gravity and its effect on light, non-Euclidean geometry and curving of space-time, impact of radio astronomy and space-age discoveries, and more. 189 b/w illustrations. xiv+250pp. 8⅜ x 9¼.   0-486-41974-6

THE VARIATIONAL PRINCIPLES OF MECHANICS, Cornelius Lanczos. Philosophic, less formalistic approach to analytical mechanics offers model of clear, scholarly exposition at graduate level with coverage of basics, calculus of variations, principle of virtual work, equations of motion, more. 418pp. 5⅜ x 8½.
0-486-65067-7